Life's Blood

Friedrich Ludwig Hünefeld
First Observer of Hemoglobin Crystals (1840) (Credit: The University of Greifswald)

Michael H. Rosove

Life's Blood

The Story of Hemoglobin

 Springer

Michael H. Rosove
Department of Medicine
Division of Hematology-Oncology
UCLA
Los Angeles, CA, USA

ISBN 978-3-031-61149-0   ISBN 978-3-031-61150-6   (eBook)
https://doi.org/10.1007/978-3-031-61150-6

This Springer imprint is published by the registered company Springer Nature Switzerland AG
The registered company address is: Gewerbestrasse 11, 6330 Cham, Switzerland

If disposing of this product, please recycle the paper.

"The world is so full of a number of things,
I'm sure we should all be as happy as kings."
A Child's Garden of Verses
Robert Louis Stevenson

Preface

Hemoglobin is the only molecule we have that captures oxygen from the air we breathe and delivers it to our tissues. My interest in it, and the study of blood in general, dates back to age ten. In that year of my life, my father died suddenly from a coronary artery blockage. He was a physician, and among his belongings at home were a microscope and a box of glass slides with smeared-out blood samples from patients labeled with their various diseases. He had prepared them during his training in the early 1930s. That year I looked with a certain wonder through the microscope at those slides and read his medical books on the diseases these patients had. My father had already introduced me to medicine and science when I was six by teaching me the anatomy of the heart. It was no surprise to my family when in college I chose to pursue medicine as a profession.

When I was a first-year medical student in 1969, Prof. Wilfried F. Mommaerts, then-Chairman of the UCLA Department of Physiology, made a comment during a lecture, something I have never forgotten: "It was a fine day in the Planning Committee when hemoglobin was invented." I understood with a quiet smile that he was talking about what evolution had miraculously wrought. Prof. Mommaerts made no secret of his admiration for this molecule—what it is, what it does, how it does it, and what can go wrong with it. I do not recall his waxing any more enthusiastically about any other molecule in his numerous lectures. His own fascination with this complex assemblage of globin proteins and heme, combined with his charisma as a lecturer, affected me greatly. His passion was one more reason beyond those boyhood glass slides why I became a hematologist and after many years came to write this book—to share with you the workings of this remarkable molecule that none of us can get by without, and the way the evolutionary "planning committee" shaped it.

Our human ancestors long ago recognized that blood is essential to life. They would have been impressed by its intense red color and sensed that blood, life, and who we are as people are inextricably linked. We see it in literature going back over three thousand years, and these connections were likely appreciated and pondered by our forebears uncountable millennia before that.

Thus from *Deuteronomy* 12:22–24—"Eat the gazelle and the deer ... But make sure you do not partake of the blood; for the blood is the life, and you must not consume the life with the flesh ... you must pour it out on the ground like water." The Hebrew words for man, earth, and red—*adam* אדם, *adamah* אדמה, and *adom* אדום—all derive from the same root word, while the word for blood—*dam* דם—derives from another, it sounds virtually the same and uses the same Hebrew letters.

e. e. cummings in his poetry collection *is 5* penned these lines: "wholly to be a fool / while Spring is in the world / my blood approves, / and kisses are a better fate / than wisdom / lady I swear by all flowers". In this lusty poem, the poet's blood stands for his passion, even his very person.

And from Johann Wolfgang von Goethe's *Faust, Part I*: "*Blut ist ein ganz besonderer Saft.*" (Blood is a very special juice.) Surrendering his soul, Faust signs his contract with the demon Mephistopheles in blood.

The pull of this "very special juice" has interested scientists for centuries; but only over the past two hundred years has investigation blossomed, thanks to advances in the scientific tools needed to find out what blood actually is. Even so, the ancient Greek physician and anatomist Erasistratus, who lived in the third century BCE and cofounded a school for research and medicine in Alexandria, came remarkably close to defining one of blood's principal functions, the physiological interplay of air and body. His research involved the then-controversial practice of necropsy, preceding by many centuries the famous dissections and illustrations of corpses during the Renaissance. He described the anatomy and function of heart valves, suggested the heart was a pump, and differentiated arteries and veins. His notions about the function of the blood vessels predated William Harvey's description of the circulatory system in the early seventeenth century by two millennia. And Erasistratus, with telling insight into physiology—which was not even a defined science at the time—suggested that arteries contained air. Never mind that the discovery of oxygen specifically had to wait to the 1770s: Erasistratus seemingly already had the idea.

Since the nineteenth century, the volume of published scientific literature concerning the nature of blood has been truly enormous involving chemists, crystallographers, physiologists, geneticists, and hematologists. All this labor has illuminated hemoglobin, one of the most fascinating molecules in all of human biology. It is the signal substance that gives blood its deep crimson hue and whose structure is uniquely designed for its vital function—the intricate "catch and release" interplay that extracts oxygen from the air we breathe, holds it long enough to deliver it to our tissues, and there lets it go to drive our metabolic functions and liberate the heat and energy without which we would not exist. It should come as no surprise that hemoglobin has long been the subject of intense study. There is no branch of physiology and medicine oblivious to its relevance.

Hemoglobin evolved so well over the course of the ages that it can recruit itself to bind oxygen, ramp up oxygen delivery instantly in response to the demand of exercising muscles, regulate vascular function, make adjustments to altitude, and transfer oxygen perfectly from a pregnant mother to her developing fetus. One could hardly ask for more. Attempts over the last three-quarters of a century to engineer a hemoglobin substitute as a treatment have thus far all fallen short despite all our technology, so complex and successful is the original design.

However, our genes programming for hemoglobin, like all genes, are susceptible to mutation. Malaria has single-handedly and dramatically changed the hemoglobin genetic landscape in areas where the disease is endemic—from sub-Saharan Africa to Southeast Asia. Malaria is caused by tiny organisms that invade red blood cells and feed on hemoglobin; numerous mutations evolved over the millennia to defend against it. They have been partly effective, but they created their own adversities in one way or another. While the most severe among them is without question sickle cell disease, others, especially the group of mutations producing what are collectively called the thalassemias, are also very common and important. Well over a quarter of people born with ancestry in the malaria belt have one or more mutations, and now with widespread dissemination of peoples all over the globe, these mutations can be found in people everywhere. Until recently, treatment options were few, but modern advances in therapeutics, especially gene therapies for the most severely affected, are proving revolutionary. Of no small interest is that hemoglobin mutations, aside from the responses to malaria, are otherwise exceptionally rare: not one has emerged as a competitor to supersede native hemoglobin in the evolutionary long view, which attests to how exceptional the basic design is.

In clinical practice, acquired (non-inherited) abnormalities of hemoglobin and red blood cells are of many kinds, they are common, and they are very often important. They require the attention of the individual's primary physician and often a hematology consultant. In addition, many readers will already be familiar with hemoglobin A1c, a mainstream, routine blood test. During the red blood cell's lifespan, glucose in the blood gradually attaches to hemoglobin, producing this variant. Measuring it is a highly useful averaging of blood sugar over many weeks for detection of prediabetes so that preventive advice can be given, and for diabetes monitoring to determine how effective medications and diet are. In another setting, that of high altitude, hemoglobin adjustments are an important part of adaptation. Hemoglobin is also susceptible to toxins that interfere with its normal oxygen transport and delivery function, none more important by virtue of sheer prevalence than carbon monoxide poisoning.

In this book, I will present the histories of ideas around oxygen, blood, and its circulation, with passing comment on alternative oxygen transporters in the animal kingdom for comparison. I will comment on human hemoglobin structure and function, how a developing fetus obtains oxygen from its mother, the magnitude of the malaria problem and the array of genetic responses that evolved to deal with it, adaptations to high altitude, blood doping, some interesting lessons from freak mutations of globins and heme, the problem of hemoglobin toxins, blood transfusion, and the ongoing pursuit of a useful hemoglobin substitute.

In a sense, this book is my personal homage to a miraculous molecule. My sincere wish is that you will come away understanding better the workings of hemoglobin and have some of the same admiration for it I have now had for over half a century.

Department of Medicine   Michael H. Rosove
Division of Hematology-Oncology
UCLA, Los Angeles, CA, USA

Acknowledgments

I express my deep appreciation and gratitude to my editor Donna Frazier Glynn for her dedication to the project, her wisdom and abundance of ideas for improvements. I am grateful to graphic designer Anthony Fernandez of the UCLA Department of Medicine for creating the graphics in Chaps. 3 and 9 and to Dr. Sophie Song, Department of UCLA Pathology and Laboratory Medicine, for the blood smear photographs in Chaps. 5 and 7. Wikipedia, Google, UpToDate, and the UCLA library system have been indispensable. I thank those who encouraged me and listened indulgently to my thoughts, dilemmas, and progress. I am grateful to all those at Springer Nature who have brought this book into production. Last but not least, AI has played no part in the research or writing, and I declare no conflict of interest of any kind.

Contents

About the Author

Michael H. Rosove is a Clinical Professor of Medicine at UCLA. He has spent almost 50 years of his career as a clinical hematologist and teacher in UCLA's Division of Hematology-Oncology.

He has an extensive list of peer-reviewed publications and has received numerous teaching awards. He founded a weekly, hour-long hematology teaching conference for fellows and faculty almost 20 years ago, a conference perpetually praised as one of the most valuable teaching experiences in the division. His original research includes being the first to establish the value of anticoagulation in preventing thrombosis and pregnancy loss in the antiphospholipid syndrome and defining the phenomenon of "decompensated erythrocytosis" in cyanotic congenital heart disease.

Dr. Rosove also has a longstanding interest in Antarctic history and bibliography and has authored numerous published works on these subjects.

Chapter 1
Blood, Body, and the "Food of Life"

Hemoglobin evolved for the purpose of transporting and delivering oxygen. Its story, in the simplest terms, entwines breath, blood, and the circulation of the blood to supply the body with the oxygen it must have for the combustion of the foods we eat to generate energy. We take our understanding of this overall process for granted, as if it has always been common knowledge. But assuredly it has not always been, even if facets of it seemed so to our ancestors predating recorded history. The written record of true scientific pursuit of these questions does reach back perhaps 2400 years, but advances were slow until a few hundred years ago. And the discovery of hemoglobin as a unique entity—when it was found accidentally as a crystallized protein—dates back not even quite two hundred years. In this chapter, we'll look at some of the seminal discoveries that preceded the discovery of hemoglobin itself.

Combustion, Respiration, and the "Food of Life"

We humans, with our curious intellects, have been driven to understand how air fuels our bodies and our lives. This inquiry has long stimulated true scientific investigation. As early as the third century BCE the Greek polymath Philo of Byzantium, who was immersed in the academic environment of Alexandria, Egypt, tried to work out the relationship between air and fire by putting a lit candle inside a closed container. He determined that there was less air in the container after the candle burned. Sixteen centuries later, Leonardo da Vinci similarly found that a portion of air was consumed in both combustion and respiration.

These observations, interesting though they were, did not clarify that air might have discrete components, and that one might be more important than another. It wasn't until the turn of the seventeenth century that a Polish scientist, Michał Sędziwój, more commonly known as Michael Sendivogius, put forward a novel

© The Author(s), under exclusive license to Springer Nature
Switzerland AG 2024
M. H. Rosove, *Life's Blood*, https://doi.org/10.1007/978-3-031-61150-6_1

idea that air was not a single substance: one of its components was what both fed flames and kept animals alive.

Around 1598, he described the heating of potassium nitrate or ammonium nitrate to release an essential life-giving gas. He did not assign a name to it, but it had to have been oxygen given the substrates he used to produce it. The discovery was all the more remarkable because Sendivogius was dealing with something one could feel as wind, or perceive in the act of breathing, but which had no visible form, smell, or taste. "Man was created of the Earth and lives by vertue of the Air," he wrote in his *A New Light of Alchemy*, "for there is in the Air a secret Food of Life ... whose invisible congealed Spirit is better then the whole Earth ... In it also is the vital Spirit of every Creature, living in all things ... [it] nourisheth them, makes them conceive, and preserveth them ... the whole structure of the World is preserved by Air ... Man dies if you take Air from him ... Nothing would grow in the World, if there were not a power of the Air."

Sendivogius's work and ideas were highly influential. His six books on alchemy were widely respected and translated from the original Latin into multiple languages, and his laboratory techniques helped establish the importance of experimentation over mere observation, or simply the assertion of an idea. Perhaps notable is that Isaac Newton, born after Sendivogius's death, had copies of Sendivogius's publications in his own library. Sendivogius's findings were later supported by the work of the renowned mid-seventeenth-century Anglo-Irish alchemist Robert Boyle who showed that a candle in a closed container evacuated of air could not burn, and a sparrow or mouse could not survive. And the great English anatomist and physiologist William Harvey, who described the circulation of blood, wrote, "... the spirit contained in the blood is as the flame in the smoke of a lamp or candle."

Later in the 1600s, John Mayow, a British chemist, conducted important investigations into respiration and the composition of air. Now that Boyle had shown that air was essential for combustion, Mayow found that it had two components, "inactive" and "active". He called the active portion "spiritus igneo-aereus" (from the Latin, a "fiery-airy spirit"), his own name for the "food of life" Sendivogius produced by burning nitrates. Mayow showed that when the element antimony was heated, it gained weight by combining with his "spiritus igneo-aereus"". He was witnessing the oxidation of antimony. The same was shown for the elements tin and lead by the seventeenth-century French physician and chemist Jean Rey.

Mayow also showed that the same gas essential for combustion was needed for respiration. He placed a candle and a small animal in a closed container. If the candle was lit, the life of the animal was shortened as the candle and the animal presumably competed for the "fiery-airy spirit". Mayow assumed therefore that the act of breathing brought into the body the essential component of air. By using bell jars inverted over water he showed that the "spirit" comprised about a fifth of air, in accordance with the now-known oxygen content of the atmosphere.

His profoundly informative work went underappreciated at the time. He had well elucidated principles for which later scientists received a great deal of credit. Scientific advances in the sixteenth and seventeenth centuries proceeded at a snail's pace compared to the extraordinary rapidity of progress today, and well over a century went by before knowledge moved forward (Figs. 1.1 and 1.2).

Fig. 1.1 Michał Sędziwój
(Credit: Hermetik
Akademie Bibliothek)

Fig. 1.2 John Mayow
(Credit: Science Source)

In the latter part of the eighteenth century, several of the finest scientific minds addressed the subject almost simultaneously. The Swedish-German pharmaceutical chemist Karl Wilhelm Scheele added substantial strength to the notion that there existed a life-supporting component of air. His experimentation caused him to conclude air had to have two components that he called "fire air" and "foul air", much as Mayow had concluded. He produced "fire air" from a number of chemical compounds including potassium nitrate, manganese dioxide mixed with phosphoric and sulfuric acids, silver carbonate, mercuric oxide, and mixing nitric and sulfuric acids. To emphasize how remarkable these experiments were, the nature and nomenclature of some of these compounds had not even been established except for their recognition by common names such as saltpetre for potassium nitrate. What Scheele had produced was oxygen, though it was not yet named. His landmark report *Chemische Abhandlung von der Luft und dem Feuer* was submitted for printing in 1775 but by a quirk in the publication process was not circulated until 1777. The English translation of Scheele's work was published as *Chemical Observations and Experiments on Air and Fire* in 1780.

In Scheele's time, scientists possessed few laboratory techniques to characterize the compounds they were studying, thus smelling and tasting toxic substances such as the salts of arsenic and heavy metals like mercury and lead were commonplace. Scheele was not left unscathed by this practice, and it is believed he died from chronic mercury toxicity, his foreshortened life of 43 years having been given to science.

Further insights came from Joseph Priestley, a prolific British chemist and educator, whose scientific work was for a time overshadowed by his political theories. Because he held unpopular views of religion and supported the American and French revolutions, mobs in Birmingham burned down his home and church in 1791, forcing him to flee; and then he escaped England altogether for Pennsylvania in 1794. But before these turmoils he conducted in 1774 a series of momentous experiments. He confirmed the findings of Sendivogius and Mayow that air is a mixture of gases and that one of them was highly reactive. He reproduced Mayow's candle-and-animal experiment, and he further discovered that placing a green plant in a sealed jar and providing sunlight restored the freshness of the air, permitting a candle to burn and a mouse to survive, a demonstration of photosynthesis, the production of oxygen by plants, although he did not recognize the role of sunlight.

Then, in his most famous experiment, performed at the Bowood House, Wiltshire, in southwest England where his laboratory has been preserved to this day, Priestley inverted a large jar over a pool of metallic mercury to seal it and then focused sunlight on the mercuric oxide he had placed within. He found the gas thus produced to be "five or six times as good as common air", that flames burned intensely within it, and that an enclosed mouse survived longer than without it. Priestley commented, when he himself breathed the unnamed gas, "The feeling of it in my lungs was not sensibly different from that of common air, but I fancied that my breast felt peculiarly light and easy for some time afterwards." (The word "fancied" is apt, since the effect of inhaling extra oxygen briefly would dissipate right away.) (Figs. 1.3, 1.4 and 1.5)

The rapidly unfolding discoveries of the decade culminated in the work of Antoine-Laurent Lavoisier, a French nobleman and chemist, who in 1772 became interested in combustion and found that phosphorus or sulfur when burned in air gained weight while consuming a portion of air. In 1774, his results with tin and lead were similar. (Later that year Priestley visited him in Paris and informed him of his own mercuric oxide experiments.) Lavoisier found that when metals combined with air, the reaction was driven not by just common air but "nothing else than the healthiest and purest part", the "eminently respirable part", since what was left

Fig. 1.3 Karl Wilhelm
Scheele (Credit:
Encyclopedia Britannica)

Fig. 1.4 Joseph Priestley
(Credit: Science History
Institute Museum and
Library)

over would not support combustion. The same year he coined the name "oxygen" for this "purest part" by contracting two Greek words to mean "acid", οξυς (*oxys*), and "to be born", γείνομαι (*genomai*), because he was struck by the fact that the combustion products of the nonmetallic elements sulfur, phosphorus, carbon, and nitrogen were acidic.

Fig. 1.5 Antoine-Laurent
Lavoisier (Private
collection, from
Jacques-Louis David,
1788)

Lavoisier's most significant contribution concerned whether combustion and respiration were different. In the winter of 1782–1783, Lavoisier, in collaboration with the great French scholar and scientist Pierre-Simon Laplace, designed a device, a calorimeter, to measure heat given off during either respiration or combustion. The outer shell of the apparatus was packed with melting snow to stabilize the temperature around an inner shell filled with ice. They measured the quantity of the exhaust gas (carbon dioxide) and heat produced by a guinea pig confined inside, then the amount of heat produced when carbon was burned to produce the same amount of exhaust. Respiratory gas exchange and combustion were found to be essentially the same, and the duality was henceforth done away with.

Lavoisier was improvidently an administrator for the *Ferme générale*, an organization that collected taxes on behalf of Louis XVI. Many among the *fermiers généraux* (farmers) became wealthy from the bonuses they received, as did Lavoisier. He used his wealth to fund his scientific research, and his philanthropy supported work in agriculture and industry. But many tax collectors were gaudily excessive in how they used their fortunes, and they were despised by the less fortunate public. Lavoisier's association would come back to haunt him in the extreme.

It so happened that among his experimental pursuits he had perfected through rigorous testing a method of improving tobacco before its sale to the public, and he promoted the French Académie des Sciences as a loyal, national interest. However, leaders of the Revolution suppressed the Académie, and Lavoisier and other tax collectors were arrested. They faced accusations of tobacco adulteration and stealing money. Lavoisier prepared their defense and denied the charges, but the judge Jean-Baptiste Coffinhal nevertheless is said to have declared, "The Republic needs neither scholars nor chemists; the course of justice cannot be delayed." Lavoisier was summarily convicted and guillotined on 8 May 1794 in Paris, at the age of 50, along with 27 co-defendants. The contemporary mathematician Joseph-Louis Lagrange, seven years Lavoisier's senior, was moved to say, "It took them only an instant to cut off that head, and a hundred years may not produce another like it." The judge Coffinhal himself was beheaded several months later. The French government exonerated Lavoisier a year and a half after his execution.

If there be a fitting epitaph to Lavoisier and his life's work, and to all those who preceded him and have followed, Lavoisier himself provided it in the preface to his *Elements of Chemistry*: "We must trust to nothing but facts: These are presented to us by Nature, and cannot deceive. We ought, in every instance, to submit our reasoning to the test of experiment, and never to search for truth but by the natural road of experiment and observation."

A long-running squabble in scientific circles ensued in the aftermath of Scheele, Priestley, and Lavoisier: Who should be credited with the discovery of oxygen? These men were contemporaries, worked independently although they were aware of each other's work, and Scheele suffered a publication delay. All their works were the culminations of significant discoveries preceding their own. In the end, the deeds of all these scientists are more important than who gets first-place accolades.

The Circulation of the Blood That Captures and Delivers the "Food of Life"

In the same general time frame that chemists were unraveling combustion, respiration, and the essential life-giving component of air, physicians and anatomists were investigating the roles of the heart, lungs, vasculature, and blood in delivering that "food of life".

In the fourth century BCE, the Greek philosopher and polymath Aristotle made note of the fact that each heart beat was followed immediately by a pulsation in the arteries throughout the body. The Greek physician and anatomist Erasistratus, who lived in the third century BCE and cofounded a school for research and medicine in Alexandria, came very close to defining one of blood's principal functions, the physiological interplay of air and body. Using the then-controversial practice of necropsy, he described the anatomy and function of heart valves, suggested the heart was a pump, and differentiated arteries and veins. With notable insight he suggested that arteries carried air.

Galen, a renowned Greco-Roman physician, surgeon, and anatomist, advanced a theory of circulation a century later. Some of his anatomical knowledge must have come as he treated the wounds of gladiators—history has it that fewer of them died after battle owing to his skill—and more knowledge came from the animal necropsies he performed. (Dissection of humans was forbidden in Rome at the time.) His theory of blood circulation, which lasted virtually unchallenged for over a thousand years, proposed that the intestines absorbed nutrients that passed in the blood through the liver and then to the heart for distribution. That was essentially correct. Also accurate was his notion that all arteries branched off from the aorta, the huge artery leaving the left ventricle of the heart. But other descriptions were incorrect. He proposed that air from the lungs enters the pulmonary veins for delivery to the left ventricle of the heart and that blood from the liver, upon reaching the right ventricle, passes through tiny holes, or pores, in the septum dividing the two sides to reach the left ventricle where the two kinds of bloods mixed. Galen reasoned that there must be pores in the septum because the major veins, the venae cavae, returning blood to the right atrium, were larger in caliber than the pulmonary artery directing blood to the lungs. (He admitted a degree of uncertainty about his explanation.[1]) From the left ventricle, he believed air-carrying blood would be distributed to the body. Galen further believed the liver was where blood was continuously being formed. These concepts were mistaken but accepted for the next eleven centuries.

Then, in the thirteenth century, an Arab born near Damascus turned Galen's dogma on its head. Ibn Al-Nafis was a brilliant and prolific physician, surgeon, anatomist, and theologian among other pursuits. After completing his medical education, he moved to Egypt at age 23, where he was recruited as chief physician at the Al-Nassri Hospital. His studies of anatomy caused him to propose a new model explaining how the body acquires air. Al-Nafis emphatically took the position that blood returning to the heart did not pass through Galen's pores in the interventricular septum because he could not find any such pores. Rather, he insisted that blood passed directly from the right side of the heart through the pulmonary artery to the lungs, where mixing of blood and air took place. Refreshed blood would then return to the left side for distribution to the body. His revelation was published in 1242 when he was 29 years old as *Sharh Tashrih al-Canun* (a transliteration from Arabic, *Commentary on Anatomy in Avicenna's Canon*). Galen's model was now superseded, but because Al-Nafis's publication was in Arabic, it remained largely unknown in the non-Arabic-speaking world until 1924, when it was chanced upon in the Staatsbibliothek zu Berlin by an Egyptian student working on his dissertation.

It was only in the sixteenth century that the understanding of circulation began to catch up with Al-Nafis, when the Dutch-Belgian physician Andreas Vesalius, widely considered the founder of modern anatomy, pursued a detailed description of the human body. Vesalius was unaware of Al-Nafis. Working in Padua, Vesalius had a steady stream of executed criminals for dissection (as sanctioned by the criminal court) in his pursuit of a detailed anatomy of the human body. At the age of 28,

---

[1] Galen, *On the Natural Functions,* III, 15.

Vesalius presented his magnificent 7-volume folio *De Humani Corporis Fabrica* (*On the Fabric of the Human Body*) with 273 finely wrought woodcut illustrations, which was published in 1543. Vesalius, like Al-Nafis, was unable to find Galen's pores; but he was reluctant to question Galen because he had no functional alternative to propose. Meanwhile, the Spanish physician Michael Servetus in 1553, also unaware of Al-Nafis, independently described the pulmonary circulation in his theological work *Christianismi Restitutio*.

The anatomists of the time were quite aware that arteries and veins were hollow and carried blood. The French anatomist Charles Estienne in 1545 described for the first time fleshy valves in veins, and these were noted again in 1574 by Hieronymus Fabricius in Padua, who described them in 1603 in his *De Venarum Ostiolis* (On the Valves of the Veins). Fabricius's description figured prominently in what would become a landmark publication by another anatomist who would follow in 1628—a comprehensive and accurate clarification of the circulation of blood.

That 1628 publication came from the 50-year-old English physician William Harvey as his monumental *Exercitatio Anatomica de Motu Cordis et Sanguinis in Animalibus* (An Anatomical Disquisition on the Motion of the Heart and Blood in Animals), published in Frankfurt. (The book is usually referred to simply as *De Motu Cordis*.) Harvey had studied medicine in Padua, received his doctorate at the University of Cambridge, and spent most of his career at St Bartholomew's Hospital, London. He was physician, lecturer, pathologist, and personal physician to King James I and then, from 1625, to the ill-fated King Charles I, to whom *De Motu Cordis* was dedicated.

In this extraordinary work, Harvey formulated a comprehensive description of the circulation based both on the observations of his predecessors and his own investigations that comprised bedside clinical observations, post-mortem examinations in humans and animals, and the barbaric dissection of living animals (both vertebrates and invertebrates).

Harvey described in detail the pumping action of the heart, its systole and diastole—the contraction and release of the muscle—and the correlation between the force of systole and the strength of the pulses. He appreciated that such factors as sleep, rest, digestion, mental activity, and exercise determined the volume of blood the heart pumped. He defined in detail that the upper chambers of the heart, the atria, contract first, sending a pulse to the lower chambers, the ventricles, with ventricular contraction following immediately. Heart valves, he explained, prevented the flow of blood from reversing; he assumed nature would not waste effort creating the valves without such a purpose. By tying off vessels or applying pressure cuffs, he showed the consequences to an extremity if blood flow through an artery were interrupted, how Estienne's and Fabricius's valves in the veins served to prevent retrograde flow, and that pressure inside arteries was greater than in veins—all strong arguments for a circulation. He described the circular motion of the systemic and pulmonary circulations. And he explained why Galen's belief that the liver generated blood was untenable. Harvey was only lacking knowledge of the capillaries in the tissues and pulmonary circulation to describe how the arterial and venous systems might be connected: he did not have the benefit of microscopy that was still in its infancy and not yet useful.[2] Even so, his explanation of the circulation of blood has only been improved upon by filling in details.

Harvey also described accurately two structures present during the fetal life of animals with lungs and their purposes. During *in utero* life, the lungs are not functional; the fetus accomplishes its respiratory needs through the umbilical vessels, placenta, and its mother. At that stage, blood bypasses the lungs by two avenues, although at the time it was unclear why it should. One is the ductus arteriosus, a vascular connection from the pulmonary artery to the aorta. Upon birth when the newborn is instantly dependent on its lungs, the ductus arteriosus (like the umbilical vessels) is no longer needed, and it begins to shrink away into scar tissue. The other is the foramen ovale, literally an oval opening permitting blood to flow from the right atrium directly to the left atrium. After birth, a membrane closes it.[3]

After a brief relocation of several years to Oxford in the 1640s during the British Civil Wars, a move made out of his concern for personal safety over his royal associations, Harvey returned to London. He spent the last 12 years of his life through 1657 in retirement despite the pleadings of colleagues to return to a professional life. A major source of his reluctance stemmed from his protracted wranglings with abusive skeptics over *De Motu Cordis* and a desire for peace and quiet in his life. His synthesis of the relationships between the heart, lungs, and circulation of the blood was eventually accepted during his lifetime, and today *De Motu Cordis* is regarded one of the most important books on human physiology ever published—a supreme example of putting forward disparate and often incongruent information from history, observation, and experimentation succinctly and logically in just 72 pages (Figs. 1.6 and 1.7).

We can only imagine that our ancestors in science, chemistry, physiology, anatomy, and medicine—Aristotle, Erasistratus, Philo of Byzantium, Galen, Al-Nafis, Vesalius, Sendivogius, Mayow, Harvey, Scheele, Priestley, Lavoisier, and others—would have been thrilled to know what we know today. Now we move toward the modern era, centuries after Harvey, when advances in technology and theory opened the door to the discovery and elucidation of the molecule that ties together blood, body, and the "food of life": hemoglobin. The next two chapters will describe its structure and function.

---

[2] The capillary bed and its purpose were discovered by means of microscopy in 1661, four years after Harvey's death, by the Italian physician Marcello Malpighi, who was born in the year *De Motu Cordis* was published.

[3] Rarely the ductus arteriosus fails to shut down. In about 20% of adults the foramen ovale does not completely close. The residual opening is usually small but may have medical consequences.

Fig. 1.6 William Harvey
(Credit: The Scientific
Revolution)

Fig. 1.7 *De Motu Cordis*
title leaf (Credit: Glasgow
University Library)

Further Reading

Al-Ghazal SK, Buairi R. Ibn Al-Nafis—the first who described the pulmonary blood circulation. J Brit Islamic Med Assoc. 2022;10(4):1–6.

[Anon.] Erasistratus. https://en.wikipedia.org/wiki/Erasistratus.

[Anon.] Karl Wilhelm Scheele. https://www.encyclopedia.com/people/science-and-technology/chemistry-biographies/karl-wilhelm-scheele.

Bejrowski P. Michał Sędziwój: an alchemist who discovered oxygen: the 456th anniversary of Michał Sędziwój's birthday. Translated by Rose A, Sirotin J. Ministry of Culture and National Heritage of the Republic of Poland; 2021.

Cameron TJ, Boyd A, Huggett MJ, et al. Phylogenomic evidence for a common ancestor of mitochondria and the SAR11 clade. Sci Rep. 2011;1:13. https://doi.org/10.1038/srep00013.

Caskey CT. Obituary: Marshall Nirenberg (1927–2010). Nature. 2010;464:44. https://doi.org/10.1038/464044a.

Galen. On the natural faculties. Brock AJ, translator. Chicago: Encyclopedia Britannica; 1952. Great Books of the Western World series, vol. 10, p. 161–215.

Gillespie CC. In: Poirier J-P. Lavoisier. University of Pennsylvania; 1996.

Harvey W. Anatomical disquisition on the motion of the heart and blood in animals. Willis R, translator of Exercitatio Anatomica De Motu Cordis et Sanguinis in Animalibus (1628). Chicago: Encyclopedia Britannica; 1952. Great Books of the Western World series, vol. 28. p. 267–304.

Keilin D. The history of cell respiration and cytochrome. Cambridge: Cambridge University; 1966.

Lavoisier A-L. Elements of chemistry: in a new systematic order, containing all the modern discoveries. Kerr R, translator of Traité Elémentaire de Chimie (1789). Chicago: Encyclopedia Britannica; 1952. Great Books of the Western World Series, vol. 45.

Mills DB, Ward LM, Jones C, et al. Oxygen requirements of the earliest animals. Proc Natl Acad Sci USA. 2014;111:4168–72.

Prager FD. Pneumatica: the first treatise on experimental physics, Western version and Eastern version. Philo of Byzantium. Wiesbaden: L. Reichert; 1974.

Priestley J. An account of further discoveries in air. Philos Trans R Soc Lond. 1775;65:384–94.

PSIBERG Team. Pauling Scale: how to use it to calculate electronegativity? https://psiberg.com/pawling-scale.

Rance P. Philo of Byzantium. In: Andrew E, et al., editors. The encyclopedia of ancient history. Hoboken: Wiley; 2021.

Rey J. Essays of Jean Rey, Doctor of Medicine, on an Enquiry into the Cause Wherefore Tin and Lead Increase in Weight on Calcination. (1630). Alembic Club Reprints. No. 11. Edinburgh: William F. Clay; 1895.

Rodwell GF. Lavoisier, Priestley, and the discovery of oxygen. Nature. 1882;27:8–11.

Sandivogius M. [sic]. A new light of Alchymy. London: A. Clark for Tho. Williams; 1674. p. 41, 97–8.

Scheele KW. Chemische Abhandlung von der Luft und dem Feuer. Uppsala: Magnus Swederus; 1777. p. 149–55.

Schofield RE. The enlightenment of Joseph Priestley: a study of his life and works from 1733 to 1773. University Park: Pennsylvania State University; 1998.

Suplee C. Joseph Priestley: discoverer of oxygen. Washington, DC: American Chemical Society; 2004.

Szydlo AZ. The life and work of Michael Sendivogius (1566–1636). Thesis for the degree of Doctor of Philosophy of the University of London; 1991. https://discovery.ucl.ac.uk/id/eprint/10124132/.

Chapter 2
The Structure of Normal Human Hemoglobin

The matter of who discovered hemoglobin is as debatable as who discovered oxygen. But we know that the understanding of hemoglobin surged forward after 1840 when a German biochemist, Friedrich Ludwig Hünefeld (illustrated on the frontispiece), pressed and dried an earthworm's blood between two glass slides and for the first time—an accidental observation, as it happened—saw bright red crystals of what he understood to be a protein. These crystals, which he also saw in dried blood from swine and humans, contained the very substance our forebears for uncountable tens or hundreds of thousands of years knew both imparted the color of blood and was integral to life itself. After Hünefeld, investigation of protein crystal structures including hemoglobin bounded forward with the hope and expectation that such detailed knowledge might yield insights on how proteins actually functioned. In this regard, hemoglobin understandably became a subject of intense interest.

Eleven years later, in 1851, the German physiologist Otto Funke extracted hemoglobin from human red blood cells and named it *blutfarbstof* (blood pigment). But it was only in 1864 that the German physiologist and biochemist Felix Hoppe-Seyler gave the name "hemoglobin" to these crystals, deriving from the Greek "hemo" for blood and "globin" meaning a ball, referring in a vague sense to the protein's overall shape. Hoppe-Seyler described the wavelengths of light hemoglobin absorbed imparting the characteristic color, and he found that oxygen associated specifically with red blood cells that contained the hemoglobin. Over the remainder of the nineteenth century, scientists proceeded to isolate and analyze hemoglobin from many animals. But still more time had to pass before a revolutionary methodology could advance knowledge of protein structures including hemoglobin.

M. H. Rosove, *Life's Blood*, https://doi.org/10.1007/978-3-031-61150-6_2

And that was X-ray diffraction crystallography. After the German physicist Wilhelm Röntgen in 1895 discovered X-rays—a form of electromagnetic radiation like visible light but with very short wavelengths—another German physicist, Max von Laue, in 1912 tried applying X-rays to copper sulfate crystals when longer wavelengths could not get through. The early part of that decade happened to be a heady time for scientists investigating atomic structure: Ernest Rutherford and others showed that atoms had a dense, positively charged core surrounded by a cloud of negatively charged subatomic particles (electrons); and Niels Bohr found that the electrons existed in defined energy states in their orbits around the nucleus but could change state and orbit by absorbing or emitting light waves of specific energies. Von Laue's X-rays, upon encountering the electrons swirling around the atomic nuclei in the copper sulfate crystal, were diffracted (scattered) in various directions. The diffraction patterns could be captured on photographic plates and reconstructed to produce three-dimensional representations of electron density, like clouds, revealing the mean positions of the atomic nuclei, chemical bonds between atoms, and the orientation of the atoms within the crystal. Von Laue detailed the mathematical principles involved; the technique lent itself to the study of innumerable chemical structures. It was revolutionary.[1] (Nowadays the tedious work of the three-dimensional reconstructions is computerized.)

In 1934, the Irish-born Cambridge scientist John Desmond Bernal and his mentee Dorothy Hodgkin applied X-ray crystallography to produce useful descriptions of various proteins, imperfect though the methods still were. Protein structures would be far more difficult to unravel than simple compounds like copper sulfate. With better methods, two Cambridge biochemists, John Kendrew in 1958 and the Austrian-born Max Perutz in 1959, through years of arduous effort, defined the structures of myoglobin and hemoglobin.[2] Perutz had worked on the hemoglobin puzzle since 1937. He and his associates later studied X-ray diffraction patterns of both oxygenated and deoxygenated hemoglobin, and in 1970 discovered something of enormous interest, that a structural shift between the two states was involved in oxygen uptake and release. This was a realization of the long-hoped-for notion that protein function might be revealed through structure (Figs. 2.1 and 2.2).

Figure 2.2 shows a model of the structure Perutz and colleagues took more than three decades to decipher. It looks like a chaotic collection of wood shavings or curled ribbon, but it is actually highly ordered.

---

[1] Röntgen and von Laue were Nobel laureates in Physics in 1901 and 1914, respectively.

[2] Perutz and Kendrew were joint Nobel laureates in Chemistry in 1962.

Fig. 2.1 Max Perutz
(Credit: Associated Press)

Fig. 2.2 A model of the
"quaternary structure" of
hemoglobin (Credit:
Worldwide Protein Data
Bank)

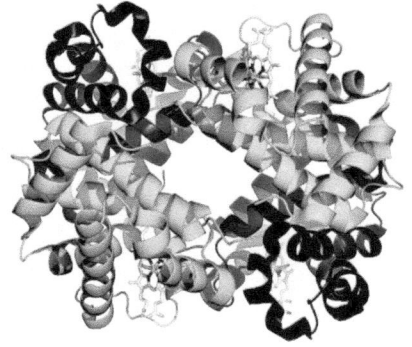

The Globin Part of Hemoglobin

Hemoglobin is a complex protein that has what is called a tetrameric structure—
meaning it is made up of four subunits. The prevailing hemoglobin in adults is made
up of two subunits labeled alpha (α), and two labeled beta (β). The amino acid
sequences of these α and β subunits were elucidated by the German physiologist
Gerhard Braunitzer in 1964. As with all proteins, the sequence of amino acids is the
"primary structure". Where positively and negatively electrically charged groups
within each chain attract each other, the amino acid strands may twist into a helix
resembling a spiral staircase, or curled ribbon, forming a "secondary structure". But

only about three-quarters of hemoglobin's subunits take on this spiral configuration: sections of the strands that contain the amino acid proline resist twisting, and at those spots, called "corners", the spirals can fold in on themselves, creating their "tertiary structure". In this folded arrangement, the α and β subunits attract each other into pairs, and two αβ pairs in turn attract each other, one facing the other head to toe, to form a tetramer, hemoglobin's final "quaternary structure". Notably, the subunits are not solidly locked together: they can move relative to each other, as Perutz described, crucial to oxygen binding and release.

While Perutz and Kendrew were teasing out the structures of hemoglobin and myoglobin—as others were doing the same with other proteins—molecular biologists were unraveling how deoxyribose nucleic acid (DNA) housed in the chromosomes transmitted its genetic code to direct protein synthesis. They identified precise stretches of DNA—the genes—that do so. Genes directing the synthesis of closely related proteins may be near each other on the same chromosome; a sequence of similar genes is a "gene region". The subunits Braunitzer elucidated are directed by the α-globin gene region on chromosome 16 (of our 23 paired chromosomes), and the β subunits by the β-globin gene region on chromosome 11.

However, more than one kind of subunit can be produced from each of the α- and β-globin gene regions. The various genes within each are ancient duplications of single ancestor genes. With time, gene duplicates variably mutated, resulting in differences in the primary structures of the subunits the genes program for. But similarities remained, referred to as homology. Of interest is that the genes are expressed in order of their position along the DNA strand as a fetus matures; the timing means they are each independently regulated. The earliest ones are vestigial remnants of our pre-human ancestry. Some no longer function and are referred to as "pseudogenes".[3]

Directed by the α-globin gene region, zeta (ζ) subunits are produced during the fetus's first trimester; they are then replaced by α subunits that will become the exclusive product of the α region for the rest of fetal and adult life. The ζ and α genes each program for a subunit of 141 amino acids. Both genes are duplicated: one of the two ζ genes became a pseudogene, while both α genes, called A1 and A2, are active.[4] Since each of the paired chromosomes have two α genes, we have four in all. The ζ and α subunits share homologous regions, but 57 of the 141 amino acids are different, strongly implying both that they diverged in the long distant past and that they function differently. Why human fetuses make ζ subunits instead of producing α subunits straightaway is unclear, but there are clues from studies in vertebrates. Human ζ subunits share even more homology with the ζ subunits in the

---

[3] The α gene region has two pseudogenes, and the β gene region has one. These ancient genes, or their regulatory genes, over time accumulated enough mutations to render the gene nonfunctional.

[4] A1 and A2 are responsible for 40% and 60% of α subunit production, respectively, as they come under different regulation. That difference is large enough to matter somewhat if one is born missing one or more α genes.

hemoglobins of chickens, mice, and rabbits than they even do with human α subunits, meaning a common and relatively preserved ancestry going back 300 million years. Even 16 of the 57 amino acid differences are found in carp, a bony fish, meaning the ancestry dates even further back, to about 400 to 500 million years. Thus ζ genes have been a constant presence in vertebrates since the classes diverged over a period of almost half a billion years, implying a strong selection pressure to preserve them. We must assume therefore that ζ subunits perform some essential purpose(s) in early human embryonic life.

The β-globin gene region has a number of genes arranged linearly along the chromosome according to chronological expression from embryonic, to fetal, to adult life, suggesting a purpose by analogy to the α-globin gene region. All the subunits programmed by the β region genes have 146 amino acids. The subunits are named in reverse alphabetical order—epsilon (ε) in earliest embryonic life, yielding to duplicated gamma subunits (^{G}γ and ^{A}γ, differing in one amino acid of no known significance), then delta (δ) and beta (β). Like the embryonic ζ gene, the embryonic ε gene had its origins about 400 to 500 million years ago. The remaining genes of the β region evolved more recently, starting about 200 million years ago. The δ and β genes are likely an ancient duplication of each other analogous to the duplicated γ genes (and the duplicated α genes). The β gene is by far more important than the δ gene as the latter is inefficient for δ subunit production: the δ gene may be on its evolutionary way to extinction as a "pseudogene" as δ subunits serve no known unique purpose; but if so, the process must be a very slow one as δ subunits are found in all primates, a class of mammals that first appeared about 60 to 90 million years ago.

Since the α region directs ζ and α, and the β region directs ε, γ, δ, and β, there can theoretically be eight different hemoglobins in humans, and seven are known to occur. As ζ gives way to α, and ε gives way to γ, four early hemoglobins[5] disappear. The principal one that emerges during fetal life is $α_2γ_2$ (HbF, commonly called either hemoglobin F or fetal hemoglobin). HbF secures the handoff of oxygen from the mother to her developing fetus, a subject to be touched upon in the next chapter. Later in the pregnancy, synthesis of $α_2β_2$ begins (HbA, hemoglobin A or adult hemoglobin). At birth, about a quarter of HbF has already been replaced by HbA, and the last hemoglobin, $α_2δ_2$ (HbA2), appears in small amounts. HbF gradually disappears during the first half year after birth, but not entirely.

Under normal circumstances—that is, with no hemoglobin mutation or problem in subunit synthesis—HbA predominates throughout life at about 96–98% of the total, HbA2 up to about 3%, and HbF up to about 1.5%. While the latter two may be remnants of our past, they are not irrelevant, as will come up later.

---

[5] During the earliest weeks of development, the embryo has four kinds of hemoglobin (Hb), mainly $ζ_2ε_2$ (Hb-Gower-1), with small amounts of $α_2ε_2$ (Hb-Gower-2), $ζ_2γ_2$ (Hb-Portland I), and traces of $ζ_2β_2$ (Hb-Portland II).

The Heme Part of Hemoglobin

The tetrameric assembly of hemoglobin just described, including all its forms from embryonic to adult, serves just one purpose—to provide a secure, sheltering home for four much smaller molecules called hemes. These are the "business centers" of hemoglobin, where the capture and release of oxygen take place. Each subunit contains one heme, and each heme houses a single iron atom that attracts a single oxygen molecule. The hemes are synthesized separately from the globin subunits, and in the final assembly of hemoglobin they are an add-on although not in the slightest a mere evolutionary afterthought.

Research elucidating and detailing the synthesis of heme involved numerous investigators during the 1940s to 1960s. Heme is made in the same cells predetermined to produce hemoglobin. A series of eight steps is involved, each governed by a unique enzyme.[6] In the last step, which takes place in the mitochondria, the iron atom is incorporated. The finished heme moiety is exported to the cytoplasm where each globin subunit secures its own heme. And so the hemoglobin molecule is now complete.

Heme is made up of four five-member rings called pyrroles, which consist of four carbon atoms and one nitrogen atom, linked via methene bridges (strong double bonds between two carbon atoms) as if holding hands. The nitrogen atoms of the four pyrroles are all oriented inward, and they can thus coordinate at the center with a single iron atom in its ferrous (Fe^{2+}) state, "like a jewel" as Max Perutz aptly described it. That is where a single molecule of oxygen (O_2) is held for transport.

Heme is a member of the porphyrin family of compounds, ubiquitous in nature and serving many functions. The basic structure is conserved in one guise or another across the biological world. In 1913, William Küster of the Institut der Technischen Hochschule in Stuttgart proposed what proved to be the correct structure; but because his complex ring assembly was previously unknown, his proposal was met with skepticism. Then in 1929 Hans Fischer of the Technische Hochschule in Munich achieved a complete chemical synthesis.[7] (In 1940 he further elucidated the structure of chlorophyll.) (Figs. 2.3 and 2.4).

In most circumstances in nature, oxygen readily takes the available electron from ferrous iron (Fe^{2+}) and converts it to the ferric (Fe^{3+}) state. Ferric oxide is rust. If that were to happen in hemoglobin, it would be rendered useless. A remarkable function of hemoglobin's tertiary and quaternary structures is each heme is protected in its own niche, while the nearby nitrogen in the pyrrole structure can share an electron to help keep iron in the ferrous state. In this way, oxygen is thus prevented from oxidizing ferrous iron: the association of oxygen to iron is kept loose, and oxygen can come and go easily. It should come as no surprise that certain relevant stretches

---

[6] The first step takes place in the mitochondria, the next four in the cell's cytoplasm, and the last three back in the mitochondria. However, all eight enzymes are programmed for by DNA in the nucleus rather than by the mitochondrion's unique DNA.

[7] Fischer was the Nobel laureate in Chemistry in 1930.

Fig. 2.3 Hans Fischer
(Credit: University of
Vienna)

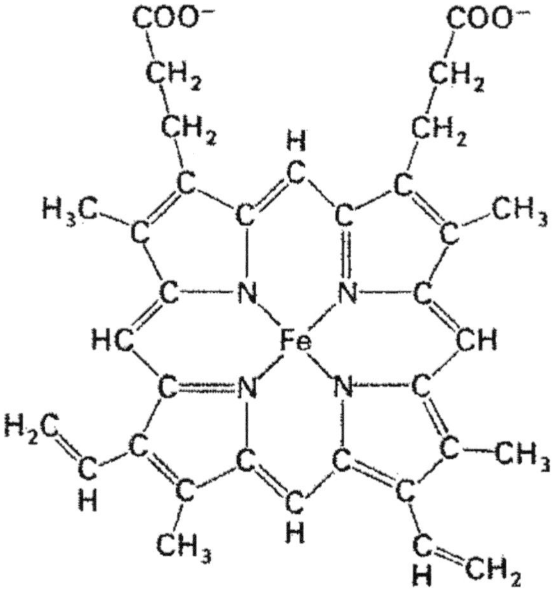

Fig. 2.4 The structure of
heme (Credit: Alchetron,
The Free Social
Encyclopedia)

of amino acids in the subunits have been highly conserved and largely unaltered throughout evolutionary history. If an accidental mutation causes a switch in any of these, hemoglobin function is impaired or crippled.

Now that we have surveyed the structure of hemoglobin, we'll next discuss what actually happens when oxygen and hemoglobin meet.

Further Reading

[Anon.] Hemoglobin. https://en.wikipedia.org/wiki/Hemoglobin.

[Anon.] Max Perutz, Father of Molecular Biology, Dies at 87. The New York Times, February 8, 2002.

[Anon.] Structural biochemistry/Hemoglobin. https://en.m.wikibooks.org/wiki/Structural_Biochemistry/Hemoglobin.

[Anon.] Technical University of Munich. https://en.wikipedia.org/wiki/Technical_University_of_Munich.

Braunitzer G, Hilse K, Rudloff V, Hilschmann N. The hemoglobins. Adv Protein Chem. 1964;19:1–71.

Clegg JB, Gagnon J. Structure of the zeta chain of human embryonic hemoglobin. Proc Natl Acad Sci USA. 1981;78:6076–80. https://doi.org/10.1073/pnas.78.10.6076.

Efstratiadis A, Posakony JW, Maniatis T, et al. The structure and evolution of the human beta-globin gene family. Cell. 1980;21:653–68. https://doi.org/10.1016/0092-8674(80)90429-8.

Fischer H, Hans O. Die Chemie des Pyrrols, 2 vols. Leipzig: Akademische Verlagsgesellschaft; 1934.

Giegé R. A historical perspective on protein crystallization from 1840 to the present day. FEBS J. 2013;280:6456–97. https://doi.org/10.1111/febs.12580.

Hardiston RC. A Cambrian origin for globin gene regulation. Blood. 2020;136:261–2. https://doi.org/10.1182/blood.2020006649.

Hoppe-Seyler F. Beiträge zur kenntniss der constitution des blutes. I. Über die oxydation im lebendem blute. Berlin: August Hirschwald; 1866. p. 133–40.

Hoppe-Seyler F. Über die chemischen und optishen eigenschaften des blutfarbstoffs. Virchows Arch. 1864;23:446–9.

Hünefeld FL. Der Chemismus in der Thierischen Organisation. Leipzig: F. A. Brockhaus; 1840.

Kendrew JC, Bodo G, Dintzis HM, Parrish RG, Wyckoff H, Phillips DC. A three-dimensional model of the myoglobin molecule obtained by x-ray analysis. Nature. 1958;181:662–6.

Küster W. Über die Konstitution des Hämins. Hoppe Seylers Z Physiol Chem. 1913;88:377–88.

London IM. Iron and heme: crucial carriers and catalysts. In: Wintrobe MM, editor. Blood pure and eloquent: a story of discovery, of people, and of ideas. New York: McGraw-Hill; 1980. p. 170–208.

Palis J, Yoder MC. Yolk-sac hematopoiesis: the first blood cells of mouse and man. Exp Hematol. 2001;29:927–36.

Perutz MF, Rossmann MG, Cullis AF, Muirhead H, Will G, North ACT. Structure of haemoglobin: a three-dimensional Fourier synthesis at 5.5-A. resolution, obtained by X-ray analysis. Nature. 1960;185(4711):416–22. https://doi.org/10.1038/185416a0.

Perutz M. Molecular anatomy, physiology, and pathology of hemoglobin. In: Stamatoyannopoulos G, et al., editors. The molecular basis of blood diseases. Philadelphia: W. B. Saunders; 1987. p. 127–78.

Pillai AS, Chandler SA, Liu Y, et al. Origin of complexity in haemoglobin evolution. Nature. 2020;581:480–5. https://doi.org/10.1038/s41586-020-2292-y.

Preyer W. Die Blutkrystalle: Untersuchungen. Jena: Mauke's Verlag; 1871.

Reichert ET, Brown AP. The differentiation and specificity of corresponding proteins and other vital substances in relation to biological classification and evolution: the crystallization of hemoglobins. Washington, DC: Carnegie Institution; 1909.

Steinberg MH, Edward JB Jr. Pathobiology of the human erythrocyte and its hemoglobins. In: Hoffman R, et al., editors. Hematology: basic principles and practice. 3rd ed. New York: Churchill Livingstone; 2000. p. 356–67.

Voon HPJ, Vadolas J. Controlling α-globin: a review of α-globin expression and its impact on β-thalassemia. Haematologica. 2008;93:1868–76. https://doi.org/10.3324/haematol.13490.

Zimmer EA, Martin SL, Beverley SM, Kan YW, Wilson AC. Rapid duplication and loss of genes coding for the alpha chains of hemoglobin. Proc Natl Acad Sci USA. 1980;77:2158–62. https://doi.org/10.1073/pnas.77.4.2158.

Chapter 3
The Function of Normal Human Hemoglobin

In the preceding chapter we looked at how hemoglobin is structured. Now let's look at its principal functions. Most important is binding oxygen in the lungs and then giving it up for metabolism in the tissues. For hemoglobin to do that, it not only must efficiently attract oxygen molecules, it must also keep its hold on them in a Goldilocks zone of "not too tight, not too loose." It does that. Miraculously. A number of factors tweak that zone. Hemoglobin also performs yet another essential task, giving the smallest blood vessels a signal to dilate and optimize blood flow. Thus, it not only carries oxygen, but also helps expedite its delivery.

How Hemoglobin Subunits "Cooperate"

A key feature of the hemoglobin molecule is that its structure allows it to transport four oxygen molecules, one per heme, instead of just one. This affords the quartet of subunits with their hemes the opportunity to act in concert to boost the tetramer's efficiency—they "cooperate" to speed the task of both binding and releasing oxygen.

Upon the arrival of even a single oxygen molecule to the heme of one hemoglobin subunit, the molecule's structures jostle, "relaxing" the entire molecule so that the remaining hemes can bind oxygen more easily. The opposite is also true: if one heme releases an oxygen molecule, the other three hemes also position themselves to release theirs. This phenomenon, called "cooperativity", is one of the molecule's most important properties. On the one hand, it allows blood entering the pulmonary capillary bed adjacent to the oxygen-rich alveoli to pick up oxygen quickly; and on the other hand, it allows the blood to efficiently unload its oxygen in the capillary beds of the tissues.

Cooperativity can be shown graphically by plotting how completely hemoglobin is oxygenated according to the amount of oxygen made available in its environment. In Fig. 3.1, the available amount of oxygen is shown on the horizontal axis, expressed

© The Author(s), under exclusive license to Springer Nature
Switzerland AG 2024
M. H. Rosove, *Life's Blood*, https://doi.org/10.1007/978-3-031-61150-6_3

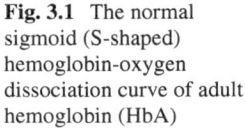

Fig. 3.1 The normal sigmoid (S-shaped) hemoglobin-oxygen dissociation curve of adult hemoglobin (HbA)

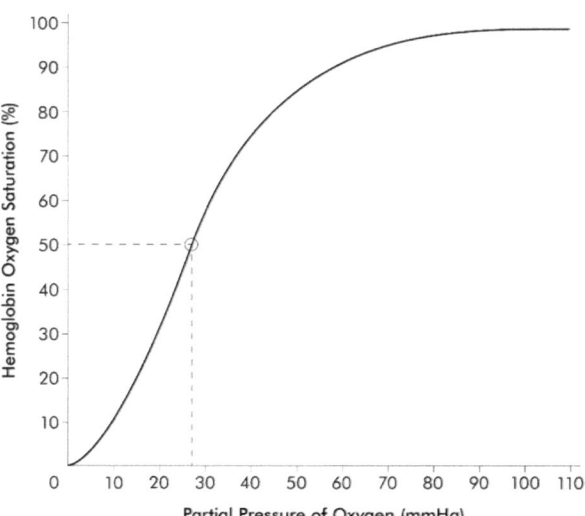

as the "partial pressure of oxygen", also referred to as "oxygen tension". All gases, whether air, oxygen, carbon dioxide, or any other, exert an outward pressure or tension, which is measured in millimeters of mercury (mmHg), just as atmospheric pressure is. The higher the number on the horizontal axis, the greater the amount of oxygen. The vertical axis shows the percentage of hemoglobin that is saturated with oxygen. A plot of hemoglobin's oxygen saturation versus oxygen tension on a graph is called the "hemoglobin-oxygen dissociation curve".

I'll get into specific numbers in a moment, but first, take a look at the overall shape of the curve (Fig. 3.1). At very low oxygen tension, hemoglobin is barely saturated. As tension increases on the graph's horizontal axis, saturation on the vertical axis gets a slow start. But then, as cooperativity kicks in, it takes less and less tension increase to catapult it to a high level as the hemes cooperatively saturate. Instead of rising gradually, the percentage of saturation shoots up, before leveling off when saturation is nearly total. The resulting curve is "sigmoid" (S-shaped), and it is not an overstatement to say that the whole of hemoglobin physiology centers around this sigmoid relationship.

To understand the curve further, both for our current discussion and because you'll see the curve and references to it again and again in this book, let's break down the numbers on the graph, starting with what exactly is the expected partial pressure of oxygen (oxygen tension), the value on the horizontal axis, in our own context.

The calculation is fairly simple, and it works like this. Mean atmospheric pressure at sea level is 760 mmHg.[1] Since air is 20.95% oxygen, the partial pressure of oxygen at sea level is 159 mmHg (20.95% of 760). When we take a breath and air arrives

---

[1] That is, millimeters of mercury, or "torr" which is virtually the same. These are the most commonly used units in the field of medicine. In other fields, 760 mm Hg may be expressed as 29.92 inches of mercury, 14.70 pounds per square inch, or 1013 hectopascals or millibars.

in our alveoli, the oxygen there is partly displaced by carbon dioxide (CO_2) being carried by the blood to the alveoli for exhalation; CO_2 has its own partial pressure of 40 mmHg. With that displacement, along with water vapor and traces of other gases, the partial pressure of oxygen in the alveoli is lower than 159 mmHg: it is about 105 mmHg. And then there is a gradient of about 10 mmHg that oxygen faces getting across the microscopic barrier between the alveoli and the pulmonary capillary blood. So the partial pressure of oxygen in blood leaving the lungs and heading for the arterial circulation in a healthy human at sea level is about 95 mmHg. Keep that figure in mind as you look at the graph, as it becomes relevant later in the book.

At that arterial oxygen tension of 95 mmHg, 98% of hemoglobin is oxygenated. When a person is at rest, mixed venous blood returning from the tissues to the lungs for reoxygenation typically has an oxygen saturation of 70% (the corresponding oxygen tension on the curve being 37 mmHg), meaning the tissues on average extracted 28% of the oxygen (98% minus 70%). Hemoglobin thus carries a usable oxygen reserve that can satisfy needs for stepped-up exercise and adverse circumstances like low blood flow for any reason, or a low level of hemoglobin for any reason. In the lungs, the difference between the alveolar oxygen tension of 105 mmHg and the usual average resting mean 37 mmHg oxygen tension of venous blood returning to the lungs is a large downhill gradient, so to speak, and hemes grab the oxygen cooperatively.

Now look at what happens on the curve when lungs malfunction (or one ascends to altitude) and oxygen tension in arterial blood falls (Fig. 3.2).[2] Follow the curve from right to left as partial pressure of oxygen declines from 95 to, say, 80, then 70, 60, and 50 mmHg (as highlighted by the dots). Up to a point, hemoglobin will continue to saturate reasonably well; but then with small further decreases in oxygen tension, saturation literally falls down the curve, and the struggle to get enough oxygen will be well underway. An oxygen saturation of about 90% in arterial blood is a useful benchmark, because it means that little reserve remains for further decreases in oxygen tension: we are on the cusp of a steep decline that physicians colloquially call "the slippery slope". Beyond that point there is a significant limitation on how much oxygen hemoglobin can pick up and deliver. It is true that there are some meaningful compensatory responses: hyperventilation can reduce the amount of CO_2 in the alveoli and thus reduce displacement of oxygen; the amount of blood the heart pumps each minute can increase; and the vasculature can dilate to improve the flow of blood. But all these have their limits.

Matters would be much worse if there were no cooperativity. Instead of a sigmoid shape, the curve would be more or less a straight angled line (not illustrated). Even if the angle were such that at an oxygen tension of 95 mmHg the hemoglobin-oxygen saturation were 100%, saturation would then be slave to perfect lung function with no tolerance for altitude, as one would immediately start falling down that angled line. The demands of the most oxygen-hungry tissues—heart muscle and exercising skeletal muscle—might well not be met.

---

[2] Lung malfunction means not being able to inhale enough air, or a disease causing oxygen not to cross the gap normally from the alveoli to the pulmonary capillaries, or a condition that causes some regions of the lung not to receive enough ventilation even though oxygen-hungry blood is flowing by.

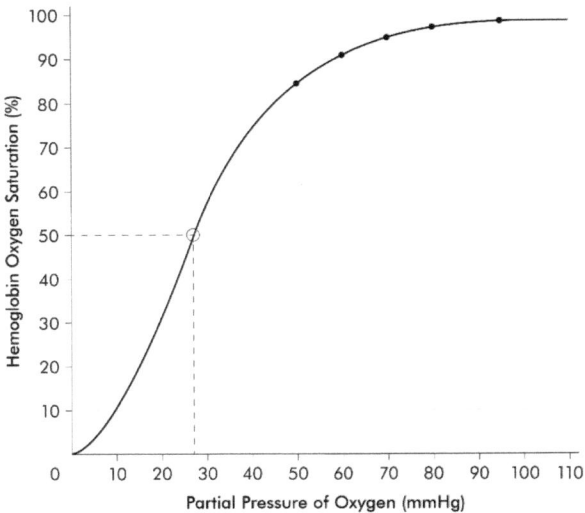

Fig. 3.2 Following the dots from right to left as the partial pressure of oxygen falls, oxygen saturation remains over 90% until it falls down the curve

Tweaking the Hemoglobin-Oxygen Dissociation Curve

The position of the curve (Figs. 3.1 and 3.2) is the average one, but a number of phenomena will affect that position. A very useful and widely employed term to describe exactly where the curve is the P50 value, the partial pressure of oxygen at which hemoglobin is 50% oxygenated. Under normal circumstances the P50 is 27 mmHg. The position of the curve can shift left or right, as shown in Fig. 3.3.

I will be referring to right- and left-shifted curves, so it's good to keep some rules of thumb in mind. When you see a rightward shift in the curve, hemoglobin does not bind oxygen as easily or hold oxygen as tightly, and it more readily gives up the oxygen that it has to the tissues, for example, to exercising muscles. However, a rightward shift also means less tolerance to lung disease or very high altitude, because the combination of less available oxygen and lower ability to bind what's there means that oxygen saturation falls down the curve sooner as arterial oxygen tension declines.

When the curve is shifted leftward, hemoglobin saturates with oxygen more avidly and holds the oxygen more tightly. Exercising muscles will not get oxygen as easily. However, a leftward shift also means greater tolerance to lung disease or very high altitude, since hemoglobin can saturate more favorably in the pulmonary circulation even as alveolar and/or pulmonary capillary oxygen tension decline (Fig. 3.3). Notice the vertical distance between the paired dots with left- and right-shifted curves. The difference is minimal at normal partial pressure of oxygen (95 mmHg), but large when it is low (50 mmHg)—the left shift is an advantage when oxygen tension is reduced.

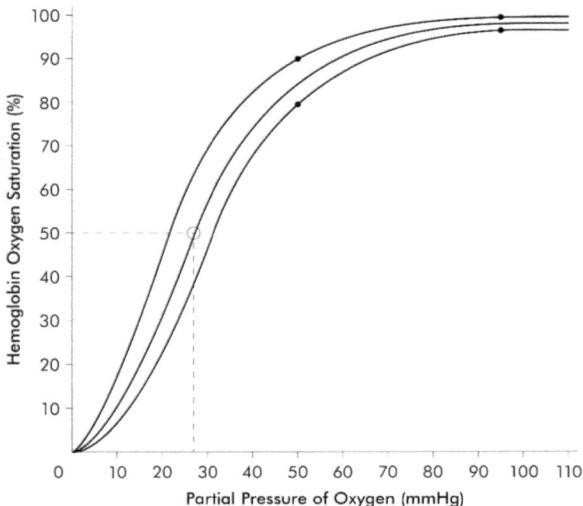

Fig. 3.3 The vertical distance between the paired dots with left- and right-shifted curves is minimal at normal partial pressure of oxygen (95 mmHg), but large at low partial pressure of oxygen (50 mmHg)

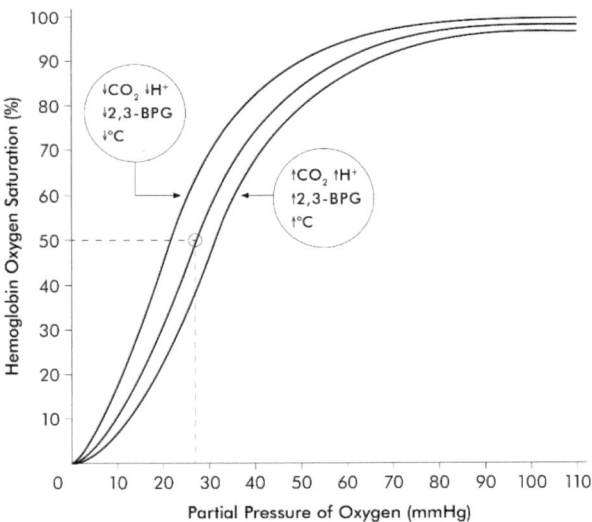

Fig. 3.4 Acidity ($\uparrow CO_2$, $\uparrow H^+$), $\uparrow 2,3$-BPG, and $\uparrow$ temperature shift the hemoglobin-oxygen dissociation curve rightward, and vice versa

Evolutionary forces had to figure out all these factors and balance a middle position; and by and large, those forces did admirably well—both through hemoglobin's genetically determined, fundamental structure, and also by three crucial factors we'll discuss that tweak the position of the hemoglobin-oxygen dissociation curve: the acidity in hemoglobin's environment (determined primarily by the amount of CO_2 produced), temperature, and 2,3-biphosphoglycerate (2,3-BPG) produced in red blood cells (Fig. 3.4).

The Effects of CO_2, Temperature, and 2,3-Biphosphoglycerate

Tissues produce CO_2 as a waste product of metabolism, and some of it is picked up by red cells. CO_2 is an acid in watery environments, generating free protons (the acid moiety) and bicarbonate. The reaction is relatively slow but is sped up in red blood cells by the enzyme carbonic anhydrase in their membranes. The protons bind to hemoglobin, reducing its affinity for oxygen and favoring release, reflected in a rightward shift of the curve. In exercising muscles, which are generating CO_2 and need oxygen, this phenomenon is particularly favorable. This adaptation was described by the Danish physiologist Christian Bohr in 1904 and to this day is still called the "Bohr effect". Bohr's own graph (Fig. 3.5) shows how the amount of CO_2 affects the curve position. It is substantial. The more CO_2 there is, the greater is the rightward shift. (Bohr happened to be one of the earliest physiologists who showed the usefulness of plotting oxygen saturation against oxygen tension graphically.)

When red cells arrive in the lungs, the relatively high oxygen tension there causes hemoglobin to give up those extra protons as oxygen binds, and the protons recombine with bicarbonate to generate CO_2 for exhalation. Our cardiopulmonary systems and neurochemical control mechanisms are strong checks and balances to maintain equilibrium.

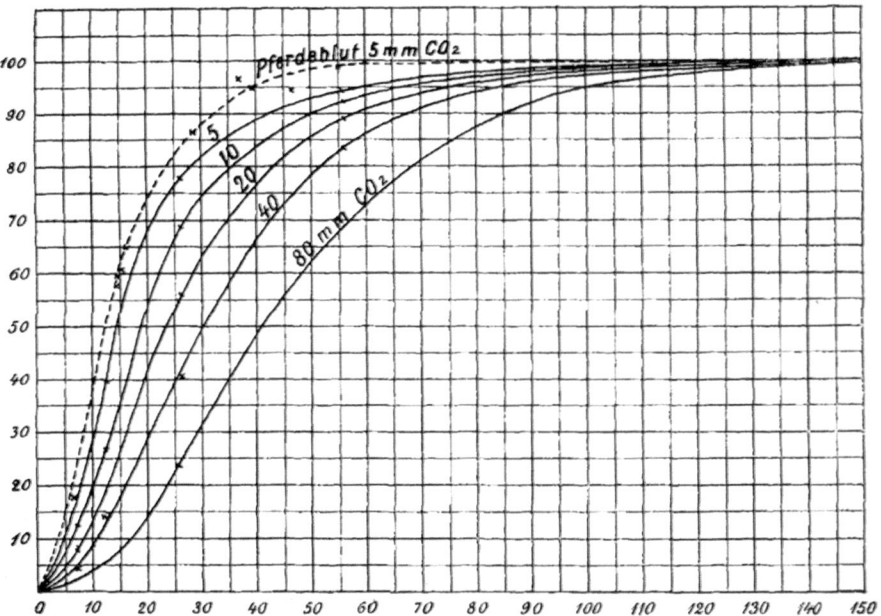

Fig. 3.5 The effect of carbon dioxide tension on the hemoglobin-oxygen dissociation curve. Reproduced from Christian Bohr's original publication (1904). (Horizontal axis, partial pressure of oxygen, mmHg. Vertical axis, % oxygen saturation)

The next factor is temperature. Higher temperature in working muscles independently drives the curve rightward, another favorable adaptation for oxygen delivery. (Hypothermia shifts the curve leftward, but the compromise in oxygen available to tissues is relatively inconsequential because hypothermia slows all metabolic functions and reduces oxygen demand.)

The third factor is 2,3-BPG. Its presence in red cells had been known since 1925, but only in 1967 was its profoundly important role in hemoglobin-oxygen interaction elucidated by Reinhold and Ruth Benesch, biochemists at Columbia University, New York.[3]

The natural tendency of hemoglobin is a left-shifted state, in which the molecule has a strong affinity for oxygen and a relative unwillingness to release it. Hemoglobin in this state is an excellent transporter but a reluctant deliverer. 2,3-BPG, a compound made almost exclusively in red blood cells, evolved to help lower hemoglobin's affinity for oxygen and allow it to more easily release it where needed. Hemoglobin's β subunits, which have positively charged "pockets", can bind negatively charged 2,3-BPG, provided the pockets are the right size. Oxygenated β subunit pockets are too small at 0.5 nm, as 2,3-BPG is about 0.9 nm. But pocket size in deoxygenated β subunits is larger at 1.1 nm, and 2,3-BPG can be admitted. That binding interferes with oxygen affinity so that hemoglobin will now not hold onto oxygen so tightly.

From the viewpoint of exercising muscles, all three of these phenomena—acidity due to CO_2 production, temperature, and 2,3-BPG, collectively conspire to have hemoglobin release oxygen more readily exactly where and when it is needed, by shifting the sigmoid curve rightward. To one extent or another, these three factors apply to all metabolically active tissues. It is difficult not to marvel at such a coordination serving the single purpose of oxygen delivery.

In addition, myoglobin, a structurally unrelated cousin of hemoglobin that also employs heme, stockpiles oxygen for muscle use. Its extremely left-shifted P50 of 2.8 mmHg means that it takes oxygen from hemoglobin and in turn makes it available for muscle metabolism (Fig. 3.6).[4]

---

[3] The two met and married in 1946 in England, he having come from Poland, and she from Berlin in 1939 via the Kindertransport. They dedicated their lifetime careers together to the study of hemoglobin.

[4] Marine mammals and birds that dive for prolonged periods foraging for food have evolved markedly elevated myoglobin content in muscle cells to store oxygen for the purpose.

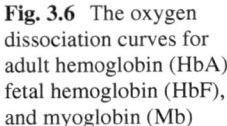
Fig. 3.6 The oxygen
dissociation curves for
adult hemoglobin (HbA),
fetal hemoglobin (HbF),
and myoglobin (Mb)

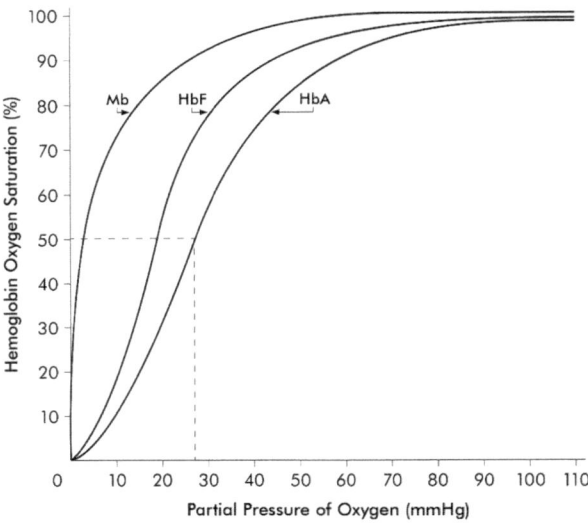

The Oxygen Handoff from Mother to Fetus

A leftward curve shift has roles to play in normal human physiology, none more
important than the transfer of oxygen from mother to fetus during pregnancy.
Because the fetus's only source of oxygen is the air breathed in by its mother, it
must have a mechanism whereby it can take oxygen off its mother's adult hemoglobin (HbA). Fetal hemoglobin (HbF) is precisely designed for the purpose because it
has greater affinity for oxygen than HbA does. Compared to HbA's P50 of 27 mmHg,
HbF's is 19 mmHg (Fig. 3.6). That left shift is explained in large part by the inherently poor 2,3-BPG binding programmed genetically into the γ subunits of HbF
($\alpha_2\gamma_2$) compared to the β subunits of HbA ($\alpha_2\beta_2$).

Thanks to the differing P50 values of HbA and HbF, oxygen flows downhill from
mother to fetus in the placenta, where maternal and fetal blood circulations come
into intimate contact. The mother is in no jeopardy as she can acquire plenty of
oxygen from the air to satisfy both her needs and that of the fetus all the way to
birth. While on the one hand it is self-evident that nature had to devise a mechanism
for maternal-fetal oxygen transfer, the HbA-to-HbF handoff and the genetic control
of fetal hemoglobin synthesis are nevertheless splendid results of evolution to humbly behold.[5] But if a prospective mother has a rare genetic condition leading either

---

[5] The fetuses or newborns of virtually all mammal species tested have relatively left-shifted hemoglobins. As in humans, most of these are due to structural differences in globin subunits, but horses
are different. Mares and their newborn foals have the same hemoglobin, but the foals are left
shifted by virtue of limited 2,3-BPG binding, a different solution to the same need.

to a very high level of persistent HbF into her adulthood, or has a markedly left-shifted mutant hemoglobin, or has received a gene therapy designed to replace the native hemoglobin with HbF, there can be important implications for her carrying a pregnancy since that downhill gradient will be diminished or absent.

After birth, the newborn, now dependent on its own lungs and no longer on its mother, has no need for HbF. The infant, just like adults, is better served by HbA that does not bind oxygen so tightly, and the baby completes the switch by the age of 6 months.

The Nitric Oxide Story

Another essential part of the hemoglobin story is nitric oxide (NO), even though it affects neither the position of the hemoglobin-oxygen dissociation curve nor, at least directly, the capture and release of oxygen in the tissues. In the 1950s, the American physiologist Arthur C. Guyton found that arterial oxygen saturation through its entire range from 100% down to 0% was a linear determinant of tissue blood flow. That is, the lower the arterial oxygen saturation, the greater the blood circulation to the tissues, a thoroughly appropriate compensation. But Guyton was unable to explain how it happened. That had to wait to the 1980s, when nitric oxide, once known only as an air pollutant, was found to be a natural product of metabolism in diverse tissues including, for our discussion, the single layer of endothelial cells lining the inside of blood vessels. Investigations that followed established NO as a short-lived, essential small molecule signal in innumerable biological systems, and found that it plays an important role in vascular physiology by inducing vessels to dilate to increase blood flow. The leaders in this field were the American researchers Robert F. Furchgott, Louis J. Ignarro, and Ferid Murad, with many others contributing.[6]

Nitric oxide is a highly reactive biological free radical because it readily donates nitrogen's free electron to a metal such as the iron in heme, or to the sulfur (thiol) atom of the amino acid cysteine. The action of NO, or its derivative nitrite (NO_2^-), which is also highly reactive, is mediated by formation of compounds called S-nitrosothiols (SNOs). In the case of hemoglobin, NO and nitrite can bind enzymatically to either heme iron or the cysteine at position 93 of the β-globin subunit ($β^{93Cys}$). It so happens that heme and $β^{93Cys}$ are functionally linked and engage in a unique choreography coordinating NO delivery to relieve resistance to blood flow in the microcirculation.

In a sense, there is a nitric oxide cycle. Hemoglobin circulating in the capillaries of the tissues delivers oxygen, and then has unoccupied hemes whose Fe^{2+} can take up NO that is generated locally in the endothelial cells. In the lungs, the high oxygen

---

[6] Furchgott, Ignarro, and Murad were joint 1998 Nobel laureates in Physiology or Medicine in 1998.

tension displaces NO off heme. NO then transfers over to β^{93Cys}, and the cysteine becomes an S-nitrosothiol (SNO). When the arterial blood reaches the microvasculature, the SNO releases the NO to dilate the microvasculature for improved blood flow as hemoglobin releases oxygen. SNO additionally aids red cells in changing shape so they can pass smoothly through the narrow capillary system. The cycle repeats itself with each tissue-to-lung-to-tissue transit, a remarkable coordination of vasodilation and oxygen delivery.

Glycated Hemoglobin

In a thoroughly different, important development, T. H. Huisman and colleagues in 1958 described variant hemoglobins that result when hexose (6-carbon) sugars attach nonenzymatically to various sites on globin subunits. The attachment of glucose in particular is permanent for the 120-day life span of the red blood cell, and while it does not alter hemoglobin function, it has proven since 1969 to be of great clinical value as a simple diagnostic blood test.

The Persian-American scientist Samuel Rahbar that year found abnormally elevated "glycated" hemoglobin (HbA1c) levels in people with diabetes mellitus, and an explosion of clinical research followed. HbA1c was found to reflect very well the average blood glucose over the previous 90–120 days. Elevated HbA1c was associated with free radical production in red blood cells, vascular inflammatory phenomena, an elevated risk of cardiovascular disease, and other well-known complications of diabetes. HbA1c is now a routine measurement in medical practice as it can detect "pre-diabetes" and help a caregiver advise a patient on optimizing lifestyle; and in frank diabetes it can determine whether control with medications and diet are adequate.[7]

The next chapter is an interlude, an opportunity to discuss some interesting hemoglobin variants and alternative oxygen transporters in the animal kingdom.

---

[7] The American Diabetes Association and American College of Endocrinology have issued guidelines concerning testing and goals of diabetes treatment. Depending on the specific laboratory's methodology, normal HbA1c is about 4.0–5.7%, borderline diabetes is 5.7–6.5%, and frank diabetes is 6.5% or greater. If diabetic control is poor, levels can run 9–12% or more. Most diabetes specialists advise treatment to maintain HbA1c 7.5% or less. Normalization has not been encouraged because of risks from over-zealous treatment. Interpretation of HbA1c assumes that red cell life span is normal or nearly so. With significant blood loss or shortened red cell survival, the average age of circulating red cells is reduced; and with less time for glycation, HbA1c will misleadingly imply better diabetic control than it actually is.

Further Reading

[Anon.] 2,3-Biphosphoglyceric acid. https://en.wikipedia.org/wiki/2,3-Bisphosphoglyceric_acid.

[Anon.] Bohr effect. https://en.wikipedia.org/wiki/Bohr_effect

[Anon.] Glycated hemoglobin. https://en.wikipedia.org/wiki/Glycated_hemoglobin.

[Anon.] Reinhold and Ruth Benesch. https://en.wikipedia.org/wiki/Reinhold_and_Ruth_Benesch.

Benesch R, Benesch RE. The effect of organic phosphates from the human erythrocyte on the allosteric properties of hemoglobin. Biochem Biophys Res Commun. 1967;26:162–7. https://doi.org/10.1016/0006-291X(67)90228-8.

Bohr CK, Hasselbalch AK. Ueber einen in biologischer Beziehung wichtigen Einfluss, den die Kohlensäurespannung des Blutes auf dessen Sauerstoffbindung übt. Skand Arch Physiol. 1904;16:401–12.

Bunn HF, Kitchen H. Hemoglobin function in the horse: the role of 2,3-diphosphoglycerate in modifying the oxygen affinity of maternal and fetal blood. Blood. 1973;42:471–9.

Gardner PR, Gardner AM, Martin LA, Salzman AL. Nitric oxide dioxygenase: an enzymic function for flavohemoglobin. Proc Natl Acad Sci USA. 1998;95:10378–83. https://doi.org/10.1073/pnas.95.18.10378.

Huisman TH, Martis EA, Dozy A. Chromatography of hemoglobin types on carboxymethylcellulose. J Lab Clin Med. 1958;52:312–27.

Ignarro LJ. Dr. NO: the discovery that led to a Nobel Prize & Viagra. Vertel; 2022.

Jensen FB. The dual roles of red blood cells in tissue oxygen delivery: oxygen carriers and regulators of local blood flow. J Exp Biol. 2009;212:3387–93. https://doi.org/10.1242/jeb.023697.

Jia L, Bonaventura C, Bonaventura J, Stamler JS. S-nitrosohaemoglobin: a dynamic activity of blood involved in vascular control. Nature. 1996;380:221–6.

Mayr E. What evolution is. New York: Basic Books, Perseus; 2001.

Michel CC. The transport of oxygen and carbon dioxide by the blood. In: Guyton AC, Widdicombe JG, editors. Respiratory physiology. Baltimore: University Park; 1974. p. 67–104.

Miyata M, Gillemans N, Hockman D, et al. An evolutionary ancient mechanism for regulation of hemoglobin expression in vertebrate red cells. Blood. 2020;136:269–78.

Peacock AJ. Oxygen at high altitude. Brit Med J. 1998;317:1063–6. https://doi.org/10.1136/bmj.317.7165.1063.

Perutz M. Reinhold Benesch (1919–1986) [obituary]. Nature. 1987;325:576. https://doi.org/10.1038/325576a0.

Perutz MF. Hemoglobin structure and respiratory transport. Sci Am. 1978;239:92–125. https://www.jstor.org/stable/24955868

Premont RT, Reynolds JD, Zhang R, Stamler JS. Role of nitric oxide carried by hemoglobin in cardiovascular physiology: developments on a three-gas respiratory cycle. Circ Res. 2020;126:129–58. https://doi.org/10.1161/CIRCRESAHA.119.31526.

Rahbar S, Blumenfeld O, Ranney HM. Studies of an unusual hemoglobin in patients with diabetes mellitus. Biochem Biophys Res Commun. 1969;36:838–43. https://doi.org/10.1016/0006-291X(69)90685-8.

Rahbar S. The discovery of glycated hemoglobin: a major event in the study of nonenzymatic chemistry in biological systems. Ann N Y Acad Sci. 2005;1043:9–19. https://doi.org/10.1196/annals.1333.002.

Ranney HM, Sharma V. Structure and function of hemoglobin. In: Beutler E, et al., editors. Williams Hematology. 6th ed. New York: McGraw-Hill; 2001. p. 345–53.

Ross JM, Fairchild HM, Weldy J, Guyton AC. Autoregulation of blood flow by oxygen lack. Am J Phys. 1962;202:21–4. https://doi.org/10.1152/ajplegacy.1962.202.1.21.

Steinberg MH, Benz EJ. Pathobiology of the human erythrocyte and its hemoglobins. In: Hoffman R, et al., editors. Hematology: basic principles and practice. 3rd ed. New York: Churchill Livingstone; 2000. p. 356–67.

Stepuro TL, Zinchuk VV. Nitric oxide effect on the hemoglobin-oxygen affinity. J Physiol Pharmacol. 2006;57:29–38.

Storz JF. Hemoglobin-oxygen affinity in high-altitude vertebrates: is there evidence for an adaptive trend? J Exp Biol. 2016;219:3190–203. https://doi.org/10.1242/jeb.127134.

Trulock EP. Arterial blood gases. In: Walker HK, Hall WD, Hurst JW, editors. Clinical methods: the history, physical, and laboratory examinations. 3rd ed. Boston: Butterworths; 1990. p. 254–7.

Wright TJ, Davis RW. Myoglobin oxygen affinity in aquatic and terrestrial birds and mammals. J Exp Biol. 2015;218:2180–9. https://doi.org/10.1242/jeb.119321.

Chapter 4
The Oxygen Transporters

Oxygen transporters are ubiquitous in nature and are found in animals, plants, fungi, and even some organisms more primitive. The transporters have one principal purpose, to deliver oxygen where it is needed for energy production. It is worth presenting them to get an overview on variations in the animal kingdom, lest we think our own hemoglobin, so important to us, is the only way nature figured out how to move oxygen around. What is remarkable is how a number of distinctly different but successful strategies *did* evolve to satisfy this essential need. Animal groups have coevolved with their oxygen transporters over a period of at least 500–750 million years.

Oxygen and Earth's Atmosphere

First, as an intro to the oxygen transporters, here are a few words about oxygen itself. As an element, it is the most prevalent in the universe after hydrogen and helium, produced by direct fusion of helium nuclei in moderately large dying stars. Its commonness is due to its nuclear stability, and it has a highly relevant feature: its strong chemical reactivity is second only to fluorine among the 98 naturally occurring elements in nature. Of enormous importance are oxygen's interactions with iron and copper, crucial in numerous physiological processes, iron in particular being right at the center of hemoglobin function, as we have seen.

At Earth's beginnings about 4.5 billion years ago, the atmosphere was mostly hydrogen and helium. Then, as the planet's crust solidified during the first 2 billion years, vulcanism released gases into the primordial atmosphere containing carbon, nitrogen, oxygen, sulfur, and others, substrates for the evolution of the earliest life forms. Only then, about 2.0–2.4 billion years ago, did molecular oxygen (O_2) begin accumulating, brought about by photosynthesizing, oxygen-producing cyanobacteria (blue-green algae) in the oceans, and perhaps also by separation of hydrogen

M. H. Rosove, *Life's Blood*, https://doi.org/10.1007/978-3-031-61150-6_4

from oxygen in water vapor by solar radiation in the upper atmosphere. This conversion in the makeup of the atmosphere was inarguably the most important biological revolution in Earth's history and is commonly referred to as the "Great Oxidation Event".

As a result, the way was cleared for the evolution of oxygen-dependent multicellular life forms. Oxygen was the promise of a more metabolically active, robust biosphere. Even so, the process got only a gradual start. Atmospheric oxygen remained below 5% because the oceans and bedrock absorbed most of it until roughly 850 million years ago. Once those sinks were saturated, atmospheric oxygen actually climbed as high as 35% around 280 million years ago, possibly from an explosion of photosynthesis in flourishing lowland swamps; but it has since vacillated downward to the current level of 20.95%. Only thanks to Earth's gravity is there an atmosphere at all: it is a very thin coating, 90% of it by mass within a mere 16 km of the surface, just 0.25% of Earth's radius.

The Eukaryotes and Their Oxygen Transporters

The eukaryotes—the most advanced living forms whose cells contain a membrane-bound nucleus containing the DNA, and which include all animals, plants, and fungi—first appeared on Earth in conjunction with the Great Oxidation Event. The power stations within eukaryotic cells are the mitochondia. Eukaryotes combust sugars and fats there with oxygen, driving the formation of adenosine triphosphate (ATP), the primary fuel of cellular metabolism. Mitochondria have their own DNA, sharing common ancestry with alphaproteobacteria, and suggesting that early in Earth's biological history, ancestors of these bacteria inserted themselves into eukaryotic cells. That may have provided them a haven for survival and replication while providing eukaryotic cells the metabolic engines on which more complex multicellular organisms would come to depend.

Unicellular and primitive multicellular animals as well as insects extract enough oxygen from their environments by simple diffusion through their outer surfaces. However, the larger the animal, the more active and metabolically demanding it is, and the more warm-blooded it is, the more diffusion will not suffice. Then a respiratory system to bring in oxygen, a specialized transporter such as hemoglobin to catch it, and a circulatory system to distribute the transporter with its oxygen are necessary. The whole is regulated by neurochemical signaling. All of these functions proceed subconsciously, as would be expected of a physiological process so integral to life.

Several very different classes of transporters evolved. All have certain characteristics in common. They are complex proteins made up of subunits. They contain either heme or a chemical structure serving an analogous function, tightly housing a metal that can exist in more than one oxidation state. Iron and copper have become over the eons the metals of choice. Their electrical ambiguity means they may act as an electron donor or recipient for the "catch and release" of oxygen. These so-called metalloproteins have existed for so long that evolution has expectedly resulted in significant differences in structure and function even within a specific type of carrier, the differences roughly proportional to the time species have gone their own divergent ways. Some transporters serve oxygen delivery; others store it; some have evolved away from or toward carrying oxygen when such is preferred; and some bind other small molecules that compete with oxygen and can be either toxic or beneficial to serve other purposes. With the exception of hemoglobin, other transporters are non-cooperative and therefore not as efficient in oxygen transport and delivery. It makes perfect sense that hemoglobin would have become the transporter of choice for the vertebrates, the most metabolically demanding of all animals. Like all proteins, the oxygen transporters each have unique sensitivity profiles to temperature, acidity, their chemical environments, and the oxygen concentration of the milieu in which they must optimally function.

Hemoglobin Variants

Hemoglobin and its variants are the most widely distributed oxygen transporters in nature. Hemoglobin employs iron; it is not the iron that confers the red color to blood but rather heme. Whether oxygen is bound affects what wavelengths of light hemoglobin absorbs. The oxygenated hemoglobin in our arteries is bright red, whereas the deoxygenated hemoglobin in veins is darker, approaching purplish maroon.

Hemoglobin is found sporadically among the major divisions of the animal kingdom that diverged from each other in the long distant past. Nowhere, however, is it such a presence as it is in the vertebrates. The vertebrate classes—jawless fishes, cartilagenous fishes, bony fishes, amphibians, reptiles, birds, and mammals—have relied on hemoglobin even as these major groups diverged from each other several hundred million years ago. Given the time frame, it should come as no surprise that hemoglobins have evolved, and some of these variants are interesting enough to be worthy of special mention.

One example is the comparative evolutionary origin of hemoglobin in the "jawless" fish—lampreys and hagfish, the most primitive vertebrate class that diverged about 500 million years ago—compared to the remaining more advanced and more recent "jawed" vertebrate classes, of which we humans are a part. It so happens there exists a globin superfamily comprising hemoglobin, myoglobin (which coopts oxygen from hemoglobin and stores it for use in muscle), and cytoglobin (with miscellaneous functions). The genes programming for hemoglobin were poorly functional in the jawless fishes but active in the rest. All classes of vertebrates, however, have cytoglobin genes. To compensate for inadequate hemoglobin gene expression in the jawless fishes, they mutated their cytoglobin genes to program for a different but functional, oxygen-transporting hemoglobin. It is an outstanding example of "convergent evolution" whereby unrelated groups of animals may independently find different solutions to fulfill the same physiological need.

There is curiously an exception to the omnipresence of hemoglobin among vertebrates. It is the jawed, bony fish family Channichthyidae, members of which are nicknamed "icefish". They lack hemoglobin entirely and are also virtually devoid of myoglobin; but they thrive well and have therefore stimulated intense scientific interest. The icefish are found only in the near-freezing Antarctic circumpolar oceans and likely evolved there about 20 million years ago when ocean currents became cold enough. Icefish do have remnant hemoglobin DNA and so represent a phasing out of need from their ancient ancestors. They have adapted by means of a very slow metabolic rate (and thus low oxygen requirement) and are benefited by the increased solubility of oxygen in very cold water, an expanded volume of blood (which is white, not red), larger blood vessels, increased cardiac output, and increased numbers of mitochondria. Particularly as the icefish are not small, ranging from about 25–50 cm in length, they have seemingly defied all the rules about larger creatures' absolute need for a dedicated oxygen transporter. In 2021 a massive colony of the icefish *Neopagetopsis ionah* was found in the Weddell Sea and made worldwide news. The colony covered an astonishing 240 square kilometers and held 260,000 nests per square km, with an estimated biomass of 60,000,000 kg (roughly equal to the weight of about 10,000 elephants).

A highly unusual use of hemoglobin is found in the giant tube worm *Riftia pachyptila*, an annelid. This worm was discovered accidentally along with numerous other species during an investigation by research submarine of hydrothermal vents in the Galápagos Rift in 1977. The worms can grow to 3 m long with diameters up to 4 cm. The worms' upper ends are red due to highly complex variant hemoglobins. Near the vents the seawater is variably rich in oxygen and carbon dioxide, but also in hydrogen sulfide (H_2S) that produces the characteristic smell of rotten eggs, sewers, and swamps. Sulfide for the vast majority of animals is toxic. However, the worms and their co-dependent bacteria living within their interiors have worked out an ingenious arrangement to benefit both. The worm's variant hemoglobin can bind sulfide reversibly as well as bind oxygen, thanks to a quirk in its evolved structure. The hemoglobin transports the sulfide to the bacteria that need it, because without hemoglobin's protection, sulfide is lost in seawater, oxidizing to forms of sulfur the bacteria cannot use. The bacteria use the sulfide, and in combination with oxygen, carbon dioxide, and various sources of essential elements, make nutrients for the worm.

Yet another unusual hemoglobin-like globin molecule is found in the large parasitic nematode *Ascaris lumbricoides* that infects an enormous number of people worldwide, especially in the tropical and subtropical regions. The worm lives in the intestinal tract and prefers as little oxygen as possible. Its variant hemoglobin binds oxygen extraordinarily tightly to make it minimally available, and then lets its nitric oxide (NO, the highly reactive free radical discussed in the last chapter) combine with the oxygen, thus disposing of it as nitrogen dioxide (NO_2).

Major mutations in hemoglobin DNA leading to significant structural changes in the protein appear not to occur often, however. In Africa, chimpanzees and bonobos split off from the line leading to humans about six million years ago, and gorillas about eight million years ago, but the amino acid sequences making up our respective hemoglobins are quite similar.

Chlorocruorin and Erythrocruorin

Some of the annelid polychaete "bristle" worms (the sabellid "feather duster", ser-pulid "organ-pipe", and chlorhaemid "green-blooded" worms) and some arthropods (other than insects) use these two transporters, which are related to hemoglobin. The names mean literally from the Greek and Latin, "red blood" and "green blood". They have structurally different heme groups that both, like hemoglobin, anchor iron. Otherwise the two kinds are quite similar. Both are highly built-up massive assemblages of subunits, conferring to them the distinction of being among the larg-est protein structures in the biological world. They have been nicknamed "giant hemoglobin". They have weaker affinity for oxygen than hemoglobin does, how-ever. As these heme-proteins are very large, they are not housed inside circulating cells like hemoglobin in the vertebrates is, but rather are free floating in the animals' "hemolymph".

Hemerythrin

Another oxygen transporter employing iron as its oxygen binder is hemerythrin. It is the conveyor in some annelid worms (the sipunculid "peanut" and priapulid "penis" worms) and some brachiopods.

The structure and function of hemerythrin differ vastly from hemoglobin. Whereas hemoglobin has a heme moiety, hemerythrin lacks such a group. Instead, it has a region of chemical structures, namely, imidazole rings of the amino acid histidine that resembles heme serving a comparable function. Without heme, hem-erythrin is not red—the oxygen-bound form is pinkish violet, while deoxygenated hemerythrin is colorless.

A major difference is how hemerythrin associates with oxygen. Whereas other oxygen transporters carry oxygen in its molecular form (O_2), hemerythrin uniquely converts oxygen to hydroperoxide.[1]

Hemocyanin

The last major oxygen transporter in the animal kingdom to mention is hemocyanin. It serves only mollusks and arthropods (other than insects) which branched off roughly together from the phylogenetic mainstream about 500 million years ago.

---

[1] Each subunit has an iron binding center anchoring two atoms of iron in reduced ferrous state (Fe^{2+}) bridged by an hydroxyl group (OH^-). An arriving O_2 is bound and converted to a hydroper-oxide ($-OOH^-$) with the hydroxyl group's positively charged proton (H^+) and two available spare electrons, one from each of the two iron atoms.

Many gastropods including the common edible snail *Helix pomatia*, found on restaurant menus as "escargot", and keyhole limpets utilize hemocyanin. It is the oxygen transporter for all the cephalopods (octopi, squid, cuttlefish, and nautiluses), some bivalves (the primitive protobranchiae), many crustaceans, some arachnids (including the somewhat related horseshoe crab), and myriapods (centipedes and relatives).

Hemocyanin is quite distinct. Like hemerythrin, it lacks a heme moiety and instead the active oxygen binding center lies in a region of imidazole rings of histidine residues. But the metal is not iron, it is copper. This departure is not off the beaten path of eukaryote metabolic processes, however, since copper, like iron, readily toggles between two natural electrical states, cuprous (Cu^{1+}) and cupric (Cu^{2+}), and is an essential element serving mitochondrial and anti-oxidant functions. One oxygen molecule associates with two copper atoms. The copper in hemocyanin imparts an unusual color—while colorless unbound to oxygen, it turns blue when oxygenated. Thus mollusks and arthropods carrying hemocyanin can truly be called "blue-blooded", as the Greek origin of the word means. The same term is still occasionally used colloquially to describe people of wealthy or noble background; but it should be obvious that that does not mean the well-heeled have hemocyanin instead of hemoglobin. And squid ink used in food recipes is not hemocyanin.

Hemocyanin is composed of subunits forming ultra-massive proteins, making hemocyanin, like erythrocrurorin and chlorocrurorin, among the largest molecules found in nature. And like erythrocrurorin and chlorocrurorin, hemocyanin is too large to be contained within blood cells and so therefore circulates free in the animals' hemolymph.

Since hemocyanin becomes saturated almost completely in low-oxygen environments and can function through a broad temperature range, it serves well the species of mollusks that have dared precarious and tenuous existences in oxygen-poor environments, the coldest, deepest marine locales, or near super-heated underwater volcanic vents.

We'll move on now to our own red blood cells, remarkable structures that house our hemoglobin.

Further Reading

Alvarez-Carreño C, Becerra A, Lazcano A. Molecular evolution of the oxygen-binding hemerythrin domain. PLoS One. 2016;11:e0157904. https://doi.org/10.1371/journal.pone.0157904.

[Anon.] Brachiopod. https://en.wikipedia.org/wiki/Brachiopod.

[Anon.] Channichthyidae. https://en.wikipedia.org/wiki/Channichthyidae.

[Anon.] Erythrocruorin. https://en.wikipedia.org/wiki/Erythrocruorin.

[Anon.] Hemocyanin. https://en.wikipedia.org/wiki/Hemocyanin.

[Anon.] Riftia pachyptila. https://en.wikipedia.org/wiki/Riftia_pachyptila.

Conant JB, Chow BF, Schoenbach EB. The oxidation of hemocyanin. J Biol Chem. 1933;101:463–73.

Dilly GF, Young CR, Lane WS, Pangilinan J, Girguis PR. Exploring the limit of metazoan thermal tolerance via comparative proteomics: thermally induced changes in protein abundance by two hydrothermal vent polychaetes. Proc R Soc Lond B Biol Sci. 2012; https://doi.org/10.1098/rspb.2012.0098.

Goodman M, Moore GW, Matsuda G. Darwinian evolution in the genealogy of haemoglobin. Nature. 1975;253:603–8. https://doi.org/10.1038/253603a0.

Lallier FH, Camus L, Chausson F, Truchot JP. Structure and function of hydrothermal vent crustacean haemocyanin: an update. Cah Biol Mar. 1998;39:313–6.

Kato S, Matsui T, Tanaka Y. Molluscan hemocyanins. Subcell Biochem. 2020;94:195–218. https://doi.org/10.1007/978-3-030-41769-7_7.

Markl J. Evolution of molluscan hemocyanin structures. Biochim Biophys Acta. 2013;1834:1840–52. https://doi.org/10.1016/j.bbapap.2013.02.020.

Minning DM, Gow AJ, Bonaventura J, Braun R, Dewhirts M, Goldberg DE, Stamler JS. Ascaris haemoglobin is a nitric oxide-activated 'deoxygenase'. Nature. 1999;401:497–502.

Purser A, Hehemann L, Boehringer L, Rogge A, Holtappels M, Wenzhoefer F. A vast icefish breeding colony discovered in the Antarctic. Curr Biol. 2022;32:842–50. https://doi.org/10.1016/j.cub.2021.12.022.

Sherman DR, Kloek AP, Krishnan BR, Guinn B, Goldberg DE. Ascaris hemoglobin gene: plant-like structure reflects the ancestral globin gene. Proc Natl Acad Sci USA. 1992;89:11696–700. https://doi.org/10.1073/pnas.89.24.11696.

Sidell B, O'Brien K. When bad things happen to good fish: the loss of hemoglobin and myoglobin expression in Antarctic icefishes. J Exp Biol. 2006;209(10):1791–802. https://doi.org/10.1242/jeb.02091.

Storz JF, Opazo JC, Hoffmann FG. Gene duplication, genome duplication, and the functional diversification of vertebrate globins. Mol Phylogenet Evol. 2013;66:469–78. https://doi.org/10.1016/j.ympev.2012.07.013.

Terwilliger NB, Terwilliger RC, Meyhöfer E, Morse MP. Bivalve hemocyanins—a comparison with other molluscan hemocyanins. Comp Biochem Physiol B Biochem Mol Biol. 1988;89:189–95. https://doi.org/10.1016/0305-0491(88)90282-9.

Weber RE, Vinogradov SN. Nonvertebrate hemoglobins: functions and molecular adaptations. Physiol Rev. 2001;81(2):569–628. https://doi.org/10.1152/physrev.2001.81.2.569.

Webster DA, Dikshit KL, Pagill KR, Stark BC. The discovery of Vitreoscilla hemoglobin and early studies on its biochemical functions, and the control of its expression, and its use in practical applications. (Review.). Microorganisms. 2021;9:1637. https://doi.org/10.3390/microorganisms9081637.

Chapter 5
The Red Blood Cell

Hemoglobin can't circulate freely on its own—that is not a physiologically viable option. It must be sequestered inside a cell dedicated to containing and preserving it: the red blood cell.

This elegant, protective solution evolved for a number of good reasons. Without an enclosure, the small hemoglobin molecule would be filtered by the kidneys into the urine, giving it a red to reddish brown color that would look like bleeding, which in a sense it would be. Losing hemoglobin every time we urinated would mean constantly having to replace it, which would require eating enough protein for their amino acids to replenish the lost globins, and getting enough iron to replace the lost hemes. There has never been a diet that could possibly keep up with such demands, even today with our high-protein diets and readily available iron supplements. We would have a perpetual deficiency of hemoglobin ("anemia", literally a "lack of blood" from the Greek), which would compromise health and function.

And then, consider that heme iron must remain in the reduced ferrous state as Fe^{2+}. The globin subunits that shelter their hemes and the adjacent nitrogen atoms making available their spare electrons to iron would be a questionable match for the constant oxidative assaults in blood threatening to convert the iron to Fe^{3+}, the ferric form, rust, incapable of transporting oxygen. But the interior of the red cell has provided a system that exists specifically to convert Fe^{3+} back to Fe^{2+}: cytochrome b5 reductase (Cyb5R), an enzyme that employs a cofactor to donate the crucial electron. The red cell's interior also produces the 2,3-BPG that is required for optimal positioning of the hemoglobin-oxygen dissociation curve. Finally, consider that hemoglobin, if it were free, would bind nitric oxide indiscriminately, interfering with normal vascular functions. For this multitude of reasons, evolution over the past half billion years has ensured that hemoglobin would be sequestered inside a cell dedicated to it. There can be no consideration of hemoglobin without taking into account the red blood cell.

M. H. Rosove, *Life's Blood*, https://doi.org/10.1007/978-3-031-61150-6_5

Fig. 5.1 Normal human red blood cells (smaller, purplish normal platelets are also visible)

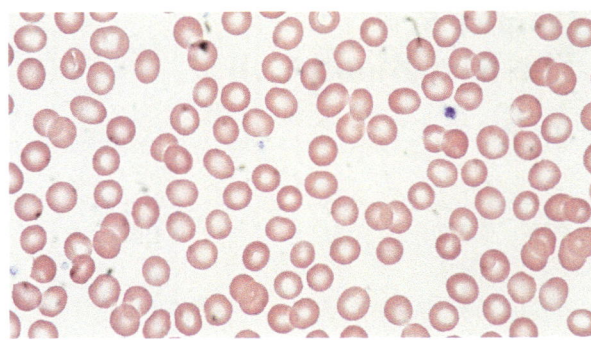

Why the Red Blood Cell is Necessary Red Blood Cell Production

Perhaps it should come as no surprise that hemoglobin is made uniquely in developing red blood cells so that it is protected from the start. Even in the first few weeks of an embryo's life, hemoglobin is contained within the "blood islands" of the tiny yolk sac. Later in the first trimester, fetal hemoglobin is made in the blood-forming cells of the liver and spleen, then continuously in the bone marrow as the fetus switches to the adult hemoglobin that will be the principal hemoglobin (with anomalous exceptions) to the end of its life. So at all stages, hemoglobins are made inside red blood cells (Fig. 5.1).

Here are a few fundamental numbers to ponder. The normal blood volume in adult humans is about 70 mL per kg body weight. That is, we are about 7% blood. An average 45% of the blood in adult men is red cells; in women, about 42%. The average size of each red cell is about 90 fL (femtoliters, 10^{-15} L), but they are not spherical—they are biconcave disks, approximately 7 μm (micrometers, one one-thousandth of a millimeter) in diameter, and about 2.0–2.5 μm thick near the circumference and 0.8–1.0 μm centrally. Red cells live in the circulation about 120 days.

Applying these numbers, an adult weighing 70 kg[1] whose blood comprises 44% red cells has about 24,000,000,000,000 (24 trillion) of them circulating at any given moment and produces about 200,000,000,000 (200 billion) red cells in the bone marrow every single day to replace the ones leaving at the end of their lifespans. These numbers, as startling and extraordinary as they may seem, are something most of us do not give a second thought. In the time you just spent reading this short paragraph, you released into the blood about 20,000,000 (20 million) new red cells.

Nearly all bone marrow is found in the centrally located bones—skull, vertebral column, sternum and ribs, pelvis, and upper arms and legs. Primitive cells in the bone marrow become committed either to producing red cells or various kinds of

---

[1] 1 kg = 2.205 lbs. 70 kg = 154 lbs.

white blood cells or platelets. All carry genetic information for producing globin chains and heme, but with the exception of the red cell line these genes get turned off. How this happens is incompletely understood, but a number of chemical signals encourage some cells to go down the red cell pathway and keep the necessary genes active.

As the earliest red cells ("erythroblasts") mature in the bone marrow, they become sensitive to, and dependent on, the hormone erythropoietin (EPO) for fullest red cell production. About 90% of EPO is made in the kidney as a response to additional chemical sensors that read the body's oxygen status and regulate how much support blood must provide for oxygen delivery. These are the hypoxia-inducible factors (HIFs) and two proteins responsible for downregulating them when oxygen is sufficient, prolyl hydroxylase domain protein 2 (PHD2) and the von Hippel-Lindau protein. Thus when there is not enough circulating hemoglobin (and hence not enough oxygen delivery), HIFs ramp up. That stimulates EPO production, which in turn stimulates the bone marrow to make more red cells. When there are now enough red cells, PHD2 and the von Hippel-Lindau protein dispose of the HIFs. That shuts off EPO production, and that in turn reduces red cell production. Much of our understanding of this oxygen sensing system is relatively recent. The American scientists William Kaelin, Jr. and Gregg L. Semenza with British scientist Peter J. Ratcliffe were the principal investigators elucidating oxygen sensing during the 1990s.[2]

Iron is an absolute requirement for heme synthesis and thus hemoglobin production. It has a fixed relationship of about 3.4 mg per gram of hemoglobin (0.34%). With iron deficiency—and there are many causes—anemia develops, and the red cell is made small on account of not enough hemoglobin to justify full size.

In 2014, another crucial protein was discovered, erythroferrone, produced in erythroblasts, and involved in the movement of iron around the body. Erythroferrone downregulates a protein made in the liver, hepcidin. That downregulation in turn permits upregulating the activity of another protein, ferroportin, which delivers iron to the circulation from two sources: dietary iron captured into cells of the small intestine, and storage iron in macrophages. The circulating protein transferrin delivers that iron to erythroblasts whose receptors for it are already revved up. In a very real sense, erythroblasts come to their own aid by directing a complex metabolic pathway outside themselves.

Developing erythroblasts rapidly generate globin subunits and heme, and hemoglobin production becomes a whopping 90% of red cell protein synthesis. Each red cell will end up carrying about 270,000,000 (270 million) molecules of hemoglobin, meaning that at any given time we are circulating a dizzying 6,500,000,000,000,000,000,000 (6.5 sextillion, or 6.5×10^{21}) molecules of hemoglobin, weighing in at about 700 g (along with 2.38 g iron), about 1% of the weight of a 70 kg adult. That is our investment in oxygen transport. The other 10% of

---

[2] Kaelin, Semenza, and Ratcliffe were jointly awarded the Nobel Prize in Physiology or Medicine in 2019.

protein synthesis comprises the components of the red cell membrane and the enzymes red cells need for their own health and function, as well as the health of the hemoglobin for which they are responsible, a remarkably low overhead investment. This entire production occurs over just six days once a multipotential marrow cell commits to the red cell pathway.

When the red cell approaches full maturity in the marrow for release into the blood, it ejects its nucleus, mitochondria, and other organelles since all its development will have been completed. Jettisoning the no-longer-needed structures allows as much room as possible for hemoglobin. That is important, because hemoglobin makes up 33% of the red cell interior, close to its limit of solubility. (At too high a concentration, hemoglobin would crystallize, become useless, and damage the red cell.) But now the red cell will live out its life only with the armaments it then possesses for its self-preservation.

During the red cell's lifespan, its hemoglobin has to continue its oxygen delivery function nonstop, and each red cell will make the lung-to-tissue-to-lung circuit every minute, roughly 170,000 times during its lifespan. Every time an oxygen molecule joins and leaves the iron atom in heme, there is a small chance the normal ferrous iron (Fe^{2+}) state will be oxidized to ferric iron (Fe^{3+}), a state of hemoglobin called methemoglobin. The aforementioned enzyme cytochrome b5 reductase (Cyb5R) and its electron-donating cofactor reduce Fe^{3+} back to Fe^{2+}. Because this process is not perfect, hemoglobin is normally about 1% methemoglobin. But this small amount has no health consequence.

The enzymes of the metabolic "glycolysis" pathway that use glucose to generate ATP are preserved in the red cell's interior. ATP provides the energy necessary to maintain the integrity of the red cell's outer membrane. It also maintains basic cellular functions including electrolyte and glucose transport and the membrane enzyme carbonic anhydrase for the favorable rightward Bohr effect. The same glucose metabolic pathway generates the 2,3-BPG to ensure a steady-state rightward position of the hemoglobin-oxygen dissociation curve. And another side pathway protects against oxidant damage, the so-called "pentose shunt" that employs glucose-6-phosphate dehydrogenase (G6PD).

An extraordinarily important property of the red cell is its deformability: its ability to change shape. The capillary bed connecting the smallest arteries to the smallest veins is where blood and tissue interface and all essential processes including gas exchange, metabolic, fluid, and electrolyte exchanges occur. Capillaries range in diameter from smaller than a red cell at 5 μm up to 10 μm.[3] Red cells, crowded as they are in blood, have to jostle and deform in order to squeeze and stampede their way through the capillary bed single file to complete their missions. There is a purpose to this microscopic bedlam, to appose as closely as possible the oxygen-rich red cells and the tissues needing that oxygen.

---

[3] For comparison, the diameter of a single scalp hair is much greater, ranging from 50 to 120 μm depending on fineness. It comes as no surprise that William Harvey without the benefit of microscopy could not see capillaries.

When the Red Blood Cell's Lifespan Is Over

Nothing lives forever, and so it is with red blood cells. But red cells die and are replaced with so little drama that the process is all but invisible. As the early twentieth-century American scientist F. Peyton Rous observed, "So subtly is normal blood destruction conducted and the remains of the cells disposed of that were it not for indirect evidence one might suppose the life of most red corpuscles to endure with that of the body." That indirect evidence was the marrow's easily recognizable red cell production activity plus the daily elimination of bile pigments identical with those attributed to the breakdown of red cells. Somewhat later, the British-born American pathologist Winifred Ashby conducted studies in which she transfused red cells to anemic patients with different but compatible blood types whose survivals could be assessed. She pinned down survival very close to the figure accepted today.

At the end of their lives, red cells are processed in the macrophages, the garbage processing cells found throughout the body. Blood platelets aid in this disposal by identifying aged red cells, attaching to them, and facilitating their identification by macrophages. Hemoglobin subunits are broken down to their constituent amino acids for recycling. Heme is enzymatically degraded into iron for reuse and a heme-like waste product, biliverdin, that is further broken down to bilirubin for bile elimi-nation and to carbon monoxide (CO).

Blood chemistry panels routinely test for bilirubin because the level is valuable in clinical diagnosis.[4] If the bilirubin level is sufficiently elevated, the skin and/or whites of the eyes may take on a yellowish tinge (jaundice). The breakdown of each heme molecule yields one CO molecule, the body's only production of it. Carbon monoxide is a waste product, but it binds avidly to heme iron in circulating red cells. Measurement of carboxyhemoglobin (COHb) is not routinely tested, but when it is, the level in non-smokers is typically under 2% and of no clinical significance. Carbon monoxide poisoning is another matter entirely, discussed later.

Of some curiosity, while red cells have presumably been hemoglobin carriers for hundreds of millions of years, they are very poorly preserved in ancient vertebrate tissue specimens. The oldest intact red cells yet identified have come from a natu-rally mummified human discovered serendipitously in the Tyrolean Alps in 1991, named Ötzi the Iceman, who died around 3230 BCE and became frozen in ice and thus preserved.[5] The first to actually see intact red blood cells was the 21-year-old Dutch biologist Jan Swammerdam in 1658, using a primitive post-Galilean micro-scope, three years before Marcello Malpighi applied microscopy to discover the

---

[4] It may accumulate because red cells are prematurely destroyed, because diseases of the liver and biliary tract have slowed down the processes that clear it away, or because a person has inherited a common, innocent enzyme deficiency of bilirubin processing.

[5] Ötzi, his artifacts, and educational exhibits are on display at the South Tyrol Museum of Archaeology, Bolzano, Italy.

capillary bed. The Italian scientist Vincenzo Menghini established the presence of iron in red cells in 1746.

Given all the foregoing, I suggest we humbly honor our red blood cells and what they accomplish. We will revisit them in upcoming chapters. The next three chapters concern the most prevalent mutations affecting human hemoglobin and the red blood cell, and the sole reason these mutations became so common: malaria.

Further Reading

[Anon.] Erythroferrone. https://en.wikipedia.org/wiki/Erythroferrone.

[Anon.] Jan Swammerdam. https://en.wikipedia.org/wiki/Jan_Swammerdam.

[Anon.] Ötzi. https://en.wikipedia.org/wiki/Ötzi.

[Anon.] Vincenzo Menghini. https://en.wikipedia.org/wiki/Vincenzo_Menghini.

Beutler E. Production and destruction of erythrocytes. In: Beutler E, et al., editors. Williams Hematology. 6th ed. New York: McGraw-Hill; 2001. p. 355–68.

Dacie JV. The life span of the red blood cell and circumstances of its premature death. In: Wintrobe MM, editor. Blood pure and eloquent: a story of discovery, of people, and of ideas. New York: McGraw-Hill; 1980. p. 211–55.

Israels LG, Israels ED. Erythropoiesis: an overview. In: Erythropoietins and erythropoiesis: molecular, cellular, preclinical, and clinical biology. Basel: Kirkhäuser; 2003. p. 3–14.

Johnson RS. How cells sense and adapt to oxygen availability. The Nobel Assembly at Karolinska Institutet. 2019. https://www.nobelprize.org/uploads/2019/10/advanced-medicineprize2019.pdf.

Ningtyas DC, Leitner F, Sohail H, Thong YL, Hicks SM, Ali S, et al. Platelets mediate the clearance of senescent red blood cells by forming prophagocytic platelet-cell complexes. Blood 2024;143:535–47.

Papyannopoulou T, Abkowitz J, D'Andrea A. Biology of erythropoiesis, erythroid differentiation, and maturation. In: Hoffman R, et al., editors. Hematology: principles and practice. 3rd ed. New York: Churchill Livingstone; 2000. p. 202–19.

Prchal JT, Gregg XT. Red cell enzymopathies. In: Hoffman R, et al., editors. Hematology: principles and practice. 3rd ed. New York: Churchill Livingstone; 2000. p. 561–76.

Semenza GL. The genomics and genetics of oxygen homeostasis. Annu Rev Genomics Hum Genet. 2020;21:183–204. https://doi.org/10.1146/annurev-genom-111119-073356.

Semenza GL. Breakthrough science: hypoxia-inducible factors, oxygen sensing, and disorders of hematopoiesis. Blood. 2022;139:2441–9.

Vallelian F, Buehler PW, Schaer DJ. Hemolysis, free hemoglobin toxicity, and scavenger protein therapeutics. Blood. 2022;140:1837–44.

Chapter 6
Malaria: Driver of Red Blood Cell and Hemoglobin Mutations

In 1990, I participated in a monthlong birding safari in Kenya, where the most virulent form of malaria is still endemic. Our group spent two days at Lake Baringo where the prevalence of malaria in the general population was especially high and where mosquitoes thrived well in the wet environment. Despite window screens, proper clothing, and insect repellant, I got about two dozen bites but did not get sick, whether by luck or thanks to the mefloquine I had taken to prevent infection. At the time, I was aware as a hematologist that the malarial organisms carried by mosquitoes infect red blood cells and feed on hemoglobin. But like most Americans I was not conscious of just how serious malaria was on a worldwide scale, or of its deep, pervasive effects on populations where disease was prevalent.

Not only is the disease one of the deadliest today, it has so threatened human health through time that it has driven the evolution of red cell and hemoglobin mutations to make blood less hospitable to the parasites—all because our immunologic defenses alone are simply inadequate. Those mutations are common in those who have ancestral ties to the malaria belt extending from sub-Saharan Africa eastward through the Middle East to Southeast Asia. The Darwinian selection pressure favoring mutations that proved purposeful has been enormous. As a result, mutations are present in a quarter of all people living today. But those mutations have their own adverse effects. If malaria had never existed, inherited abnormalities of hemoglobin and red cells would be freak occurrences—they are, in fact, exceptionally rare otherwise, mostly one-off events, so successfully evolved are normal red cells and hemoglobin as we know them.

I will devote the next two chapters to the mutations themselves, but to put them in context, we'll first talk about malaria itself, which joins them at the root.

M. H. Rosove, *Life's Blood*, https://doi.org/10.1007/978-3-031-61150-6_6

The Disease

In the minds of many people who live outside the malaria belt, two key words about malaria are mosquito and fever. A mosquito bites, the victim is infected, gets a fever, and the fever is treated. With luck, that may be the extent of it. But not always. A more accurate way to think of the disease is as an invasion of the blood. In the broadest terms, to be detailed later, it looks like this: A mosquito bites, malaria parasites enter the blood stream and travel to the liver. There they multiple and mature, move out, and invade red blood cells. Inside the red cells, they multiply, then burst out along with their toxic waste products while damaging the red cells. The cycle self-perpetuates. Immune defenses are easily overwhelmed. In the worst case scenario, the parasites cause red cells to stick to small blood vessels and block them, leading to organ damage. With that, the disease is devastating; and it is particularly hard on children. It is not just a fever. Unrecognized and untreated, it can be a disaster.

Humans have been dealt the dreadful consequences of malaria since the beginning of the Holocene epoch, pegged at 11,700 years ago with the end of the last ice age and retreat of the glaciers. The rise of agriculture and civilizations coincides precisely with the rise of malaria which has since caused more disease and death than any other infections except tuberculosis and smallpox. That might seem surprising in light of such pandemics as influenza in 1918 which killed about 50 million people worldwide, HIV/AIDS that since 1979 has killed about 40 million, SARS-CoV-2 which has been responsible for about 7 million deaths just since 2020, and the plague epidemics of Europe and England—but it is a fact. Smallpox, which first appeared about 3500 years ago, resulted in about 300 million deaths in the twentieth century alone until it was eradicated.[1] Tuberculosis, which appeared in humans about 6000 years ago, killed 1.6 million people in 2021 according to the WHO.

Malaria, meanwhile, was responsible for 241 million cases and 627,000 deaths in just 2020 according to the WHO. In the twentieth century, it killed between 150 and 300 million people. Ninety-five percent of malaria deaths occur in sub-Saharan Africa, and young children there are disproportionately over-represented among the victims. Despite advances in mosquito control and disease treatment, progress in the region has stalled, as the WHO *World Malaria Report 2021* states that of 38 countries (mostly sub-Saharan), death rates had not improved in 14 and had worsened in the rest.

---

[1] Smallpox, without question historically one of mankind's most horrific tormentors, was declared eliminated worldwide by the World Health Organization (WHO) in 1980 after relentless immunization programs. It is the only human infection eradicated at large. The virus has no animal reservoir but still exists in two (hopefully) tightly controlled and secured research laboratories, one at the Centers for Disease Control and Prevention (CDC) in Atlanta, and the other at the Vector Institute in Koltsovo, Russia. Concerns will therefore remain—namely, laboratory accident, bioterrorism, or biological warfare—unless both agree to destroy their stockpiles.

Malaria has been carried by humans outside its original geographical confines, but too recently for new mutations to have evolved in new locations. The disease was carried from Africa to Europe as far north as current England and Denmark in the first century CE by merchants, colonists, and Roman soldiers. It was unknown in the Western Hemisphere until it was introduced by European settlers who brought slaves from West Africa starting in the late fifteenth century. The disease was a menace in the United States during the nineteenth century, when about a million Americans were infected, even as far west as California in 1849 during the gold rush. The disease had a substantial impact on the prosecution and outcome of a number of battles during the Civil War of 1861–1865. Six American presidents contracted malaria strictly within the continental United States: George Washington, James Monroe, Andrew Jackson, Abraham Lincoln, Ulysses S. Grant, and James A. Garfield. Two more, Theodore Roosevelt and John F. Kennedy, were infected abroad. During World War II, malaria was a serious problem for American soldiers in the Pacific theater and was the principal reason for establishing the CDC right after the war. Malaria had a huge impact on American infantry forces during the Korean War (1950–1953) and the Vietnam War (1964–1973). The recent description of several malaria cases in Florida and Texas is a reminder that reemergence at home in the U.S. always remains.

The parasites responsible are single-cell protozoans of the genus *Plasmodium.* Five species infect humans. *P. falciparum* is responsible for three-quarters or more of cases in sub-Saharan Africa and almost exclusively everywhere for life-threatening and fatal cases. *P. vivax* causes most of the rest in Africa and a higher percentage outside Africa, including the Western Hemisphere.[2]

Malaria ranges in clinical severity from latent all the way to explosive with life-threatening manifestations. Generally there is a strong correlation between the level of parasites in the blood and the severity of disease. Early diagnosis and treatment may prevent the worst outcomes. However, symptoms in mild cases, which include fever, fatigue, nausea, headache, abdominal pain, and muscle aches, are not specific, and physicians are likely to consider a number of infections and non-infectious illnesses as they come to a diagnosis.

While diagnosis of malaria is straightforward, suspicion of it is the high hurdle, because without suspicion the very simple, best diagnostic test might not be performed. And that is examination of an appropriately stained blood smear (a drop of blood spread out on a glass slide) by an experienced hematologist or laboratory technician with an excellent, high-resolution microscope. In the United States with only about 2000 cases annually (virtually all acquired abroad), examining a blood smear might not be thought of. That is unfortunate, because plasmodia inside red

---

[2] Occasional cases are due to *P. ovale, P. malariae,* or *P. knowlesi. P. vivax* and *P. ovale* typically cause fever on alternate days, "tertian malaria", counting the first day as day 1, with fever repeating on day 3 and so on. *P. malariae* causes "quartan malaria", with fever every 3 days (days 1 and 4). These intermittent fever patterns were recognized in ancient medical writings long before the cause was known. Fevers in falciparum malaria are different, usually chaotic.

cells are easy to recognize, and the appearances differ between species.[3] Given that malaria remains extremely common across the tropical and subtropical regions, and that people easily travel and relocate, malaria remains a relevant concern for healthcare providers everywhere in the world today. A Nigerian native, for example, may board plane flights for Los Angeles with symptoms and arrive at LAX a day or two later in a state of shock.

Clues specific for malaria are recent presence or residence in a malaria zone, or fever every second or third day. Symptoms first appear about 7–30 days after the original hit by the "hot" mosquito. Severe, life-threatening cases are additionally marked by rapid red cell destruction with anemia and free hemoglobin passing through the kidneys into the urine, discoloring it red to red-brown ("blackwater fever"), as well as jaundice from an increased blood level of bilirubin from processed heme. Red cells adhering to blood vessel lining cells may obstruct the vessels and cause "cerebral malaria" (altered cognition, coma, or seizures), and lung or kidney injury. Those who survive not infrequently have permanent disabilities. Even mild cases pose serious concern in pregnancy as maternal malaria may result in the death of the fetus, preterm labor, low birth weight, or death of the newborn.

First Discoveries

Little was known about malaria until the late 1870s when the French physician Charles Louis Alfonse Laveran, in his early 30s and working in Algeria, saw abnormalities inside red cells in a patient who died from malaria. He reported the findings along with illustrations to the Société médicale des hôpitaux de Paris (Medical Society of Paris Hospitals) late in 1880 and published them the following year. He suggested that a microscopic infectious entity involving red cells was responsible. His conclusion was questioned for several years until refined blood staining and microscopy permitted confirmation. This was the first time anyone had found that a protozoan could cause human disease. Laveran had an honored academic career and established a laboratory for the study of tropical diseases at the Pasteur Institute in Paris (Fig. 6.1).[4]

The prevailing hypothesis at the time was that malaria was an airborne condition. In fact, the name "malaria" derives from the Italian, "bad air". But then in 1895, the 38-year-old Scottish physician Ronald Ross working in Calcutta discovered a plasmodium species in the gastrointestinal tract of an anopheles mosquito. (He contracted malaria himself the next year.) His results, published in 1897, established the

---

[3] Other tests include rapid antigen, host antibody, and polymerase chain reaction (PCR) tests, fluorescent microscopy, and detection of hemozoin. However, use of these is tempered by availability, variable sensitivity and specificity, level of expertise needed in interpretation, time delays to reporting results, and expense.

[4] Laveran was the Nobel laureate in Physiology or Medicine in 1907.

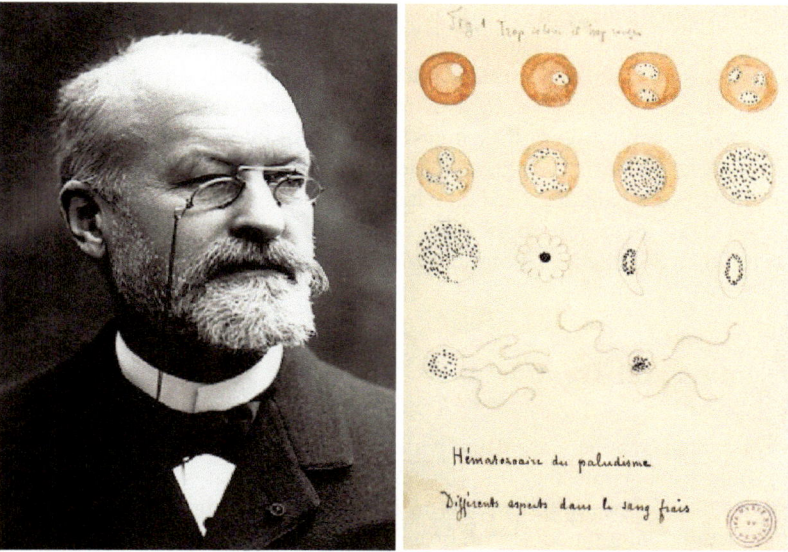

Fig. 6.1 Charles Laveran and his drawings of *P. falciparum* (Credit: Nobel Foundation archive [left], and Musée du service du santé des armées au Val-de-Grâce [right])

role of the mosquito in transmitting disease. He detailed the life cycle of plasmodia in avian malaria.[5]

In the years since mosquitoes and plasmodia were linked, the plasmodium life cycle has been investigated extensively. A hungry female anopheles mosquito—the males feed on nectar and not blood—bites an already infected human for a blood meal, most often at dusk or during the night. Infected blood contains plasmodia at various stages of its life cycle, including "gametocytes", the male and female sexual forms of the parasite, which are ready for sexual consummation. They mate within the mosquito to produce an extravaganza of "sporozoites" that migrate to the mosquito's salivary glands. When the mosquito, once again hungry, now bites an uninfected person, the bite injects sporozoites that are taken up by the bitten person's liver cells. There they will asexually generate thousands of more advanced, invasive "merozoites" over a period of about 2–10 days. The liver cells then rupture, releasing the myriad merozoites into the circulation where they now insert themselves into red cells. (The liver phase is not serious enough to produce symptoms, or abnormal blood tests of liver function.) The merozoites continue to replicate asexually in red cells, rupturing and releasing merozoites to infect other uninvolved red cells. Some of the merozoites from red cells become sexually potent gametocytes. A hungry, uninfected female mosquito on a quest for a blood meal bites, ingests the gametocytes, and begins the cycle anew. (*P. vivax* and *P. ovale* sporozoites may produce separate "hypnozoites", from the Greek meaning sleeping, stashing

---

[5] Ross was the Nobel laureate in Physiology or Medicine in 1902.

themselves quietly away in cells of the liver, spleen, or bone marrow, only to reactivate months or even years later.)

Laveran, and later many others, described an unusual "malarial pigment" in infected red cells originally named "hemozoin", a name that has stuck. When merozoites feed on hemoglobin, they digest the protein into amino acids for their own use in protein synthesis. The heme moiety is toxic to them, however, and it must be disposed of. We do not degrade heme inside red cells, thus plasmodia must take that job upon themselves, which they do, converting heme to an insoluble crystalline waste product, β-hematin. This is hemozoin, a useful diagnostic clue on blood smears.

Prevention and Treatment

Prevention and treatment are equal partners in lessening the impact of malaria. It is impossible to overstate the importance of mosquito abatement and prevention of mosquito bites.[6] Plasmodia depend on a confluence of high mosquito density and high human population density to provide a self-perpetuating reservoir of infective organisms. Abatement means simply eliminating standing water where mosquitoes lay their eggs in the same areas where significant numbers of people live.[7] Preventing bites means using common, low-tech barriers out of doors like hats, netting, and repellants—the CDC provides an updated list. Indoors, measures are window screens and netting around sleeping beds. A person visiting an area with significant risk of mosquito-borne disease should consider these easy measures. If traveling into a malaria zone, prophylactic drug treatment starting before arrival is necessary.

But preventive measures alone have fallen short for centuries, fueling efforts to treat the disease that reach back almost two millennia to China. Ge Hong in the year 340 CE assembled a compendium of herbal medicines. In 1596, Li Shezhen, a wellspring of knowledge in many fields, devoted much of his life to cataloguing herbal remedies and presciently used qinghaosu tea to treat symptoms of malaria. The Chinese scientist and pharmacist Tu Youyou, with formal training in herbal medicine and working in Beijing, spent years looking for an antimalarial from about two hundred candidate botanicals before finding one in the 1970s: qinghaosu (artemisinin), which she extracted from *Artemisia annua* (sweet wormwood), the same plant that had been used to treat fevers for centuries by herbalists. Artemisinin now is one of the most important treatments for falciparum malaria, and is believed to have

---

[6] In this regard, the same principles applying to malaria apply to other regional mosquito-borne infections, including West Nile virus, Eastern equine encephalitis virus, Zika virus, Chikungunya virus, and Dengue fever.

[7] Mosquitoes take advantage of freestanding water in ditches, discarded tires, buckets, planters, stagnant birdbaths, flowerpots, septic tanks, neglected fountains and pools, and improperly covered water storage containers and reservoirs. These gifts to mosquitoes should be eliminated, and judicious spraying with larvicides remains an option.

saved millions of lives.[8] In the U.S., the oral artemisin derivative artemether in combination with lumefantrine, both FDA-approved, is the treatment of choice for uncomplicated falciparum malaria. For severe cases, another FDA-approved artemisin derivative, intravenous artesunate is first-line treatment, often given in combination with other anti-malarials. The triumph of artemisinin has recently been a bit lessened by the emergence in East Africa of partially resistant *P. falciparum* strains.

Until the twentieth century, East and West were minimally in communication over medical ideas and discoveries; advances therefore evolved largely independently. Starting in the seventeenth century, the most commonly used and well-known treatment for malarial fever in the West was quinine, extracted from the bark of the cinchona tree from the Andean slopes of Peru. Likely apocryphal, a legend says that a native in the Andean jungle afflicted with fever was cured by drinking bitter water from a stagnant pool among quina-quina trees. His fellow natives then used bark extracts from the tree for fever successfully. Legend further relates that the Spanish Countess of Chinchón while in Peru had fever cured by the bark. Irrespective of the unconfirmable parts of the story, it is a fact that the countess brought bark back to Europe. The bark's active compound was isolated in 1820 by the French chemists Pierre-Joseph Pelletier and Joseph Bienaimé Caventou and named quinine after the local name for the tree. It became the effective treatment for malaria until better derivatives were introduced over a century later, notably chloroquine in 1934, and later primaquine, mefloquine, and tafenoquine.

Today, a number of drugs are used for prevention and treatment.[9] Decisions are governed by which plasmodium species is under consideration, whether plasmodia in a given area have acquired drug resistance, and whether the individual is an acceptable risk for certain side effects, or is pregnant or planning pregnancy. Of note is that the CDC has warned about the serious worldwide problem of counterfeit and substandard drugs.

Newest therapies are now focused on two kinds of immune treatments: monoclonal antibodies provide the body passive immunity, and vaccines stimulate one's own antibody formation. The development of any effective immune-promoting therapy has been extraordinarily frustrating and difficult, but contenders have emerged. Two monoclonal antibodies against *P. falciparum* have been investigated, named CIS43LS and L9LS, the latter a redesign of the former with greater potency. In two small early studies, Gaudinski and colleagues in August 2021 using CIS43LS, and Wu and colleagues in August 2022 using L9LS, in studies approved and funded by the National Institute of Allergy and Infectious Diseases, reported highly encouraging results. Some volunteers received antibodies, others untreated served as controls, and all were deliberately bitten by mosquitoes carrying *P. falciparum*. The

---

[8] Tu was a Nobel laureate in Physiology or Medicine in 2015.

[9] These include chloroquine, primaquine, mefloquine, tafenoquine, doxycyline, atovaquone/proguanil, and artemisinin alone or in combinations. Dapsone-pyrimethamine, sulfadoxine-pyrimethamine with amodiaquine are used for prevention in Africa. Quinine and some analogs are no longer recommended owing to greater toxicity. Primaquine or tafenoquine is typically used to clear the hypnozoites of *P. vivax* and *P. ovale*.

antibodies caused no notable safety issues, and they proved almost perfectly preventive. (All study subjects were followed closely, and those who became infected, virtually all of them controls, were treated successfully with standard drugs.)

An expanded study of CIS43LS, also under the auspices of the National Institute of Allergy and Infectious Diseases, authored by Kayentao and colleagues and published in November 2022, involved 330 individuals in Mali with a single dose of antibody during a 6-month seasonal period of high falciparum malaria risk. All subjects were cleared beforehand of possible *P. falciparum* with drugs and then equally divided between two dose levels of intravenous CIS43LS or placebo. Again, treatment was effective. Large studies are currently underway with L9LS.

A malaria vaccine called RTS,S/AS01 entered clinical trials in 2015 in African countries. Investigators enrolled 15,459 children, and falciparum malaria was reduced by 36% over 4 years in children aged 5–17 months; benefits in those younger were unclear. In areas where falciparum malaria is prevalent, vaccination was estimated to prevent 484 deaths per 100,000 children. In 2021, a University of Oxford study on another vaccine R21/Matrix-M was found 77% effective over 12 months' follow up.

In October 2021, the WHO recommended large-scale use of RTS,S/AS01 (trade name, Mosquirix) for use in children to prevent death from falciparum malaria, the first such recommendation for a protozoan parasite. Close to a million children in Ghana, Malawi, and Kenya have been vaccinated so far. The vaccine requires four doses, and current vaccine production meets only a small percentage of need. The aggregate cost of production and deployment is very high: financial and technical support have come from the Bill and Melinda Gates Foundation, the Malaria Vaccine Initiative, the Vaccine Alliance, and Glaxo-Smith-Kline.

Taken together, improved mosquito abatement, prevention of mosquito bites, and advances in drugs, monoclonal antibodies, and vaccines are poised to have an enormous beneficial impact on reducing the tolls from malaria. Total eradication may no longer be a pipe dream. But even if malaria were to disappear entirely, its mark on human health would endure in the form of genetic changes in the red cells and in hemoglobin itself that evolved to protect against the disease. These changes would be passed on from generation to generation for centuries and millennia into the future although their prevalence would likely decline.

We'll examine in the next two chapters the immense imprint malaria has had on human hemoglobin and red cell evolution, and the corresponding fallout.

Further Reading

Achan J, Ambrose OT, Erhart A, et al. Quinine, an old anti-malarial drug in a modern world: role in the treatment of malaria. Malar J. 2011;10:144. https://doi.org/10.1186/1475-2875-10-144.
Alonso PL, O'Brien KL. A malaria vaccine for Africa—an important step in a century-long quest. N Engl J Med. 2022;386:1005–7. https://doi.org/10.1056/NEJMp2116591.

[Anon.] Charles Louis Alfonse Laveran. https://en.wikipedia.org/wiki/Charles_Louis_Alphonse_Laveran.

[Anon.] Hemozoin. https://en.wikipedia.org/wiki/Hemozoin.

[Anon.] Li Shizhen. https://en.wikipedia.org/wiki/Li_Shizhen.

[Anon.] Malaria. https://en.wikipedia.org/wiki/Malaria.

[Anon.] Malaria. https://www.cdc.gov/malaria/about/disease.html.

[Anon.] Malaria. https://www.cdc.gov/malaria/diagnosis_treatment/diagnostic_tools.html.

[Anon.]. Nobel lectures, physiology or medicine 1901–1921. Amsterdam: Elsevier; 1967.

[Anon.] Ronald Ross. https://en.wikipedia.org/wiki/Ronald_Ross.

[Anon.] https://www.nobelprize.org/prizes/medicine/2015/press-release/.

[Anon.] https://entomologytoday.org/2014/02/17/at-least-eight-u-s-presidents-had-malaria/.

Arrow KJ, Panosian C, Gelband H, editors. Saving lives, buying time: economics of malaria drugs in an age of resistance. In: Institute of Medicine (US) Committee on the Economics of Antimalarial Drugs. Washington, DC: National Academies; 2004.

Carter R, Mendis KN. Evolutionary and historical aspects of the burden of malaria. Clin Microbiol Rev. 2002;15:564–94.

Daily JP. Monoclonal antibodies—a different approach to combat malaria. N Engl J Med. 2022;387:460–1. https://doi.org/10.1056/NEJMe2207865.

D'Alessandro U. Monoclonal antibodies against malaria. N Engl J Med. 2022;387:1898–9. https://doi.org/10.1056/NEJNMe2213148.

Datoo MS, Natama HM, Somé A, et al. Efficacy and immunogenicity of R21/Matrix-M vaccine against clinical malaria after 2 years' follow-up in children in Burkina Faso: a phase 1/2b randomised controlled trial. Lancet Infect Dis. 2022;22(12):1728–36. https://doi.org/10.1016/S1473-3099(22)00442-X.

Gaudinski MR, Berkowitz NM, Idris AH, et al. A monoclonal antibody for malaria prevention. N Engl J Med. 2021;385:803–14. https://doi.org/10.1056/NEJMoa2034031.

Kayentao K, Ongoiba A, Preston AC, et al. Safety and efficacy of a monoclonal antibody against malaria in Mali. N Engl J Med. 2022;387:1833–42. https://doi.org/10.1056/NEJMoa2206966.

Laurens MB. RTS,S/AS01 vaccine (Mosquirix™): an overview. Hum Vaccin Immunother. 2020;16:480–9. https://doi.org/10.1080/21645515.2019.1669415.

Mihreteab S, Platon L, Berhane A, et al. Increasing prevalence of artemisinin-resistant HRP2-negative malaria in Eritrea. N Engl J Med. 2023;389:1191–202. https://doi.org/10.1056/NEJMoa2210956.

Nadjm B, Behrens RH. Malaria: an update for physicians. Inf Dis Clin North Am. 2012;26:243–59. https://doi.org/10.1016/j.idc.2012.03.010.

Nussenzweig RS, Vanderberg J, Most H, Orton C. Protective immunity produced by the injection of X-irradiated sporozoites of Plasmodium berghei. Nature. 1967;216:160–2. https://doi.org/10.1038/216160a0.

Pelletier [Pierre Joseph] et [Joseph Bienaimé] Caventou. Des recherches chimiques sur les Quinquinas. Annales de Chimie et de Physique 1820;15:289–318, 337–65.

Ryan ET, Succi ME, Paras ML, Klontz EH. Case 4-2024: a 39-year-old man with fever and headache after international travel. N Engl J Med. 2024;390:549–56. https://doi.org/10.1056/NEJMcpc2309382.

Sato S. Plasmodium—a brief introduction to the parasites causing human malaria and their basic biology. J Physiol Anthropol. 2021;40:1. https://doi.org/10.1186/s40101-020-00251-9.

Tangpukdee N, Duangdee C, Wilairatana P, Krudsood S. Malaria diagnosis: a brief review. Korean J Parasitol. 2009;47:93–102. https://doi.org/10.3347/kjp.2009.47.2.93.

Thomson MC, Stanberry LR. Climate change and vectorborne diseases. N Engl J Med. 2022;387:1969–78. https://doi.org/10.1056/NEJMra2200092.

Tu Y. The discovery of artemisinin (qinghaosu) and gifts from Chinese medicine. Nat Med. 2011;17:1217–20. https://doi.org/10.1038/nm.2471.

Wells T, Donini C. The science behind the study: monoclonal antibodies for malaria. N Engl J
    Med. 2022;387:462–5. https://doi.org/10.1056/NEJMe2208131.
World Health Organization. Guidelines for the Treatment of Malaria. 2nd ed; 2010.
World Health Organization. https://www.who.int/news-room/fact-sheets/detail/malaria.
World Health Organization. World Malaria Report 2021. Switzerland: WHO; 2021.
Wu RL, Idris AH, Berkowitz M. Low-dose subcutaneous or intravenous monoclonal antibody to
    prevent malaria. N Engl J Med. 2022;387:397–406. https://doi.org/10.1056/NEJM0a2203067.

Chapter 7
Hemoglobin and Red Blood Cells Versus Malaria

The plasmodial parasites that cause malaria put themselves largely outside the reach of the immune system by sequestering themselves in red cells, liver cells, and sometimes cells in the spleen and bone marrow. That reality has forced an entirely different kind of defense strategy over the millennia: evolutionary changes in the genetic programming of hemoglobin and red cells. The Darwinian selection pressure favoring mutations offering any protection at all has been enormous. The result is that two billion people—a quarter of all people living today—carry mutations.

The advantages the mutations have brought about, however, have been counterbalanced by problems they have created. Every change to hemoglobin or the red cell to resist malaria has compromised their structure or function in one way or another. Yet because malaria is so likely to cause serious debility or death, and that falciparum malaria in particular kills so many young children especially in sub-Saharan Africa, evolution has favored mutations to improve the likelihood children will survive to childbearing age, which, in a sense, is just DNA looking after itself. In the case of the mutation causing sickle cell disease, the trade-off is a terrible tragedy. Similar trade-offs pertain to one degree or another throughout the malaria belt with the host of other mutations.

All human populations have been paying a price for the modest, incomplete protection these mutations provide. If malaria had never existed, inherited abnormalities of hemoglobin and red cells would be freak occurrences—they are, in fact, exceptionally rare otherwise, mostly one-off events, so successfully evolved are normal red cells and hemoglobin as we know them. But malaria has been our reality, and individuals with family ancestries in malaria zones may have more than just one mutation. Given the mobility people have in the modern world, these mutations can be found almost anywhere.

In this chapter, we'll look at the non-sickle conditions. Sickle cell disease will get a chapter of its own.

M. H. Rosove, *Life's Blood*, https://doi.org/10.1007/978-3-031-61150-6_7

Some Background on DNA, mRNA, and Protein Synthesis

To appreciate and understand red cell and hemoglobin variants, it's helpful to review some background on DNA. In 1944, Oswald Avery, a Canadian-American molecular biologist, settled an important prevailing question whether deoxyribose nucleic acid (DNA), housed in the chromosomes, was the carrier of the genetic code. It was. DNAs were linear strings of four kinds of nucleotides, adenine (A), guanine (G), cytosine (C), and thymine (T). In 1951, the British X-ray crystallographer Rosalind Franklin suggested DNA's helical structure. Then, in 1953, James Watson and Francis Crick formally proposed what was the correct structure of DNA, a double helix, two identical DNA strands oriented antiparallel to each other, with chemical bridges like ladder rungs connecting paired nucleotides on each side of the helix.[1] In their landmark paper in *Nature*, the two men recognized the structure's chief significance—that it lent itself to unwinding and replication. (They mentioned Franklin only in passing despite her essential contribution.)

The DNA sequences in a strand are secured by strong "covalent" bonds between successive nucleotides to preserve what they program for. Each nucleotide has a chemically pre-ordained partner in the complementary strand, A with T, and C with G. A nucleotide and its partner are a "base pair". The bridges between the strands are looser "hydrogen" bonds that permit easy unwinding for replication when a cell divides. As the strands unwind, the cell manufactures a new, precisely matching partner strand for each of the originals that will be available to form a new double helix with another matched strand. The human genome has about 3,000,000,000 (three billion) base pairs.

Every second we produce millions of new cells, each requiring precise duplication of those unwound DNA strands. But there is no such thing as perfect DNA replication. Reasons have been proffered why not perfect—unusual nucleotide spatial orientation, transient proton shifts, and others. But simply put, the process makes mistakes. A mismatch of nucleotides within a base pair is a common error. Sometimes extra nucleotides are added to the strand (insertions), or intended nucleotides never get copied (deletions). Sometimes segments of DNA in different chromosomes swap positions and form new functional genes or leave behind gaps. Some viruses are capable of inserting their own DNA into ours. And radiation or toxins may alter the DNA.

"Errors" by their very name imply they are "wrong", but the reality is that they are a fact of unintentional biochemistry. Errors in the germ cells that produce ova and sperm are the foundation of evolution when they happen to program for improved adaptation and survival (always coincidentally and unintentionally). If germ cell DNA replication were perfect, we would never have evolved even as far as cyanobacteria. All living creatures exist as they are only because "errors" have yielded new ideas and designs in nature fit for propagation. An "error" in a

---

[1] Watson, Crick, and Maurice Wilkins were joint Nobel laureates in Physiology or Medicine in 1962. Many believe that disregarding Rosalind Franklin was an injustice.

surviving germ cell may accumulate in the population over generations if it provides a survival advantage and can thereby become a new norm.

DNA replication makes a mistake about once in every 100,000 nucleotides. As there are 3 billion base pairs, that means 30,000 mistakes each time a cell divides. Proofreading and error correction enzymes, in a sense "DNA spellcheckers", take care of nearly all of them, but that still leaves some. When a mutation occurs in a non-germ cell line *and* confers a growth advantage to the affected cell and its progeny, *and especially* if there are also mutations impairing DNA proofreading and repair, the consequence may be cancer. Some germ cell mutations may similarly predispose to malignancy. A reality is that the same principles of DNA replication that have driven evolution over the eons and created the miraculous profusion of life on the planet are the same that drive the development of malignancies, or protection against malaria: from DNA's own viewpoint, it is merely "survival of the fittest DNA". All is blameless biochemistry.

In addition to providing the genetic blueprint for the next generation of cells, DNA has another central role—directing the synthesis of proteins. In the 1950s, Marshall Nirenberg, Har Gobind Khorana, Robert Holley, and others figured out how DNA's string of nucleotides directed assembly of 20 different kinds of amino acids to produce a protein's primary structure.[2] Unwound DNA is first transcribed in the cell's nucleus to messenger ribonucleic acid (mRNA), a replica of DNA but differing in its sugar moiety, and with uracil (U) substituting for thymine. The mRNA is exported from the cell's nucleus to the cytoplasm where it binds to ribosomes, the cell's protein manufacturing plants. A sequence of three mRNA nucleotides, a "codon", programs for a specific amino acid. With four possible nucleotides in each of the three positions, 64 codons are possible. Sixty-one of them program for specific amino acids (with a fair number of duplications), and three codons tell the ribosomes when to stop adding amino acids to the growing chain.[3]

Mutations and their amino acid substitutions and consequences will come up as we delve into the red cell and hemoglobin variants.

Red Blood Cell Mutations

Among red cell variants, one stands out prominently as a defense against malaria: deficiency of the enzyme glucose-6-phosphate dehydrogenase (G6PD), which is integrally involved in preventing oxidative red cell and hemoglobin damage. The most persuasive evidence that the deficiency evolved as a malaria response is the high prevalence of G6PD deficiency in native peoples of the malaria belt and the rarity of the deficiency elsewhere. In addition, G6PD-deficient red cells infected with *P. falciparum* appear to be more rapidly cleared by macrophages. G6PD

---

[2] Nirenberg, Khorana, and Holley were joint Nobel laureates in Physiology or Medicine in 1968.

[3] About 2% of our DNA is "coding", programming for about 20,000 to 25,000 proteins. The rest is "non-coding" with various functions, or representing remnants of old genes no longer needed and mutated toward, or already arrived at, non-function.

deficiency states are almost without exception explained by single nucleotide substitutions affecting just one amino acid. Well over two hundred of them have been described, but only two are common. The "Mediterranean variant" has G6PD enzyme activity less than 5% of normal, and the consequences are significant. About 10% of Africans have the so-called "A– variant" with 10–23% activity, and the implications are milder.

Most individuals with the Mediterranean type remain well, however, despite the fact that some individuals have perpetual low-grade shortening of red cell survival, until they inadvertently have a meal of fava beans—also called broad beans, ironically a staple food from the Mediterranean eastward, and common ingredient in falafel—or are exposed to an oxidizing drug, or have a serious infection. Then they are in for a bad though thankfully brief surprise, which fortunately is seldom life-threatening. Fava beans are delicious, but they contain two compounds that are converted in the gastrointestinal tract and in the beans themselves into the chemicals divicine and isouramil, noxious to red cells if their G6PD defense is wanting. Abrupt red cell and hemoglobin injury produces variably severe fatigue, abdominal pain, anemia, jaundice, and methemoglobin production. Recovery is usually prompt over the ensuing days.

A seeming irony is that three drugs useful in preventing and/or treating malaria—primaquine, tafenoquine, and dapsone—cannot be used in G6PD-deficient individuals because they cause oxidant damage; therefore G6PD testing is mandatory before any of these (and several other highly offensive drugs) are used.

As the G6PD gene is located on the X chromosome and all men are XY, all red cells in a male carrying a G6PD mutation are deficient. As women are XX, and because one X in each red cell precursor extinguishes itself at random, women with one mutation produce two populations of red cells, one deficient and the other normal. With two abnormal genes, all the red cells are deficient.

Genetic abnormalities in red cell membrane proteins may also protect against malaria as the malarial parasites gain entry to red cells first by attaching themselves to specific protein receptors in the red cell membrane. Duffy is the receptor for *P. vivax*, and MNS, Gerbich, the red cell surface antigen CD44 in conjunction with other membrane proteins actuate *P. falciparum* invasion. Mutations of these may theoretically impair the entry of plasmodia.

Hemoglobin Mutations

Far more significant than red cell mutations in terms of numbers of people affected and the implications are hemoglobin mutations that evolved to protect against malaria. Anomalies fall into two categories, *unbalanced production* of α and β subunits (the thalassemias) and *intrinsic abnormalities* of β subunits (the hemoglobin variants).

The Thalassemias

It is not going too far to state that serious thalassemia syndromes are the price populations pay for the malaria protection that on balance benefit mainly those who have mild forms with few or no symptoms.

Under normal circumstances, even numbers of α and β subunits are produced by remarkably well-coordinated expression of the respective α- and β-globin gene regions, despite the fact they are on the separate respective chromosomes 16 and 11. A mutation that impairs production of either α or β subunits results in an imbalance. Imbalance has several consequences: the total number of normal hemoglobin tetramers that can be produced in each red cell is reduced; reduced hemoglobin leads to anemia, which triggers erythropoietin production to expand red cell production; and erythroblasts produce more erythroferrone, which leads to over-absorption of iron from the diet; and surplus subunits may complex with themselves into tetramers that are functionless for oxygen delivery and damage the red cell. Collectively, these disorders are an important group of very common disorders referred to as the thalassemias.

"Thalassemia", the name coined by the renowned American physician George Whipple in 1932, is a contraction of the Greek word *thálassa* (θάλασσα, referring to the Mediterranean Sea, a region where the condition is common) and *haima* (αἵμα, meaning blood). Research into these disorders during the past century has been voluminous and driven in large part by pure intellectual pursuit, because for many years treatments were limited to blood transfusion, iron chelation to rid the excess iron from red cell transfusions, family counseling, and little else. But now, rapid advances in drug treatments, gene therapies, and gene editing have dramatically excited the field of therapeutics and given practical value to what might once have been viewed simply as academic pastimes.

Like G6PD deficiency, the thalassemias occur almost exclusively in populations long resident in the malaria belt. About 250–400 million people have a thalassemic disorder, about 3–5% of the world's population. Given the dispersion of descendants from ancestral malaria regions throughout the world, physicians everywhere, particularly primary care physicians and hematologists, encounter thalassemic patients regularly.

In all the thalassemias, red cells are smaller than normal because of deficient hemoglobin production, and blood smears reveal misshapen and distinctive "target" red cells that on average are paler than normal from a reduced level of hemoglobin within them. Mild cases are without symptoms. In more severe cases, the spleen may be enlarged making red cells; and in the rarest, most severe cases, an intense drive to produce red cells may enlarge the liver, expand marrow activity deforming bones not normally committed to blood cell production, and produce masses of marrow tissue elsewhere. If a person receives many red cell transfusions over time, iron overload ensues, causing organ damage and a predilection to certain bacterial infections. However, the very vast majority of cases are on the mild end of this spectrum, and severity in a given individual rarely worsens over a lifetime.

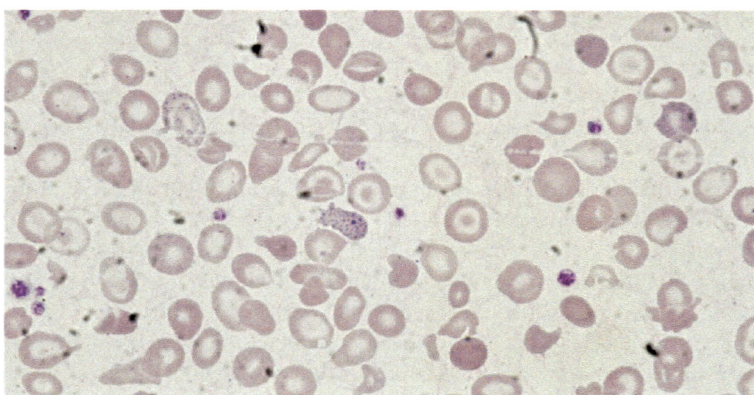

Fig. 7.1 Blood smear from a patient with β-thalassemia intermedia. Numerous misshapen, pale, and "target" red cells are present; smaller purplish platelets are also present

Thalassemias are commonly gauged according to three degrees of increasing clinical consequence or severity—thalassemia minor, thalassemia intermedia, and thalassemia major. Another common classification is whether the anemia is, or is not, severe enough to require occasional or regular red cell transfusion. These two kinds of descriptions do not mention the background genetics. In 1959, V. M. Ingram and A. O. Stretton segregated the thalassemias into α and β thalassemias based on which globin subunit is deficient. All of these classifications are informative, and together they remain bedrock descriptions (Fig. 7.1).

Readily available blood tests describe whether someone has anemia, what the mean red cell size is, and which kinds of hemoglobins are present. Expensive genetic panels can define most (but not all) mutations, and whole genome sequencing can pick up others. Such tests are often used in thalassemia intermedia and thalassemia major cases, and in people with thalassemia minor planning to have children when genetic counseling is advisable. Those who have thalassemia themselves and those who treat it will be particularly interested in the details that follow concerning the thalassemia types.

The alpha thalassemias. All functional hemoglobins have four globin subunits, two each programmed from the α- and β-globin gene cluster regions. Mutations in α genes are common across the entire malaria belt; the frequency in some areas exceeds 15–20%. As the α gene is duplicated (A1 and A2) on each chromosome 16, we have four of them, perhaps an indication of their critical importance, since α subunits are the backbone of all hemoglobins after the disappearance of ζ subunits in early embryonic life. It is possible to be missing anywhere from one to four α genes. Most of the mutations are deletional leaving no subunit production, but others leave some intact, in which case the severity of the thalassemia is less significant.

If *one* α gene is missing (expressed as –α/αα, each side of the slash representing the inheritance from one or the other parent), the result is only a slight lowering of red cell size with no anemia, and the condition is most often unnoticed. If *two* are missing, ($-\alpha/-\alpha$) or ($--/\alpha\alpha$), then red cells are unmistakably small with borderline or mild anemia (α-thalassemia minor). If *three* are missing ($--/-\alpha$), significant anemia is constant, red cell survival is shortened, and red cell transfusion may be necessary (α-thalassemia intermedia). The unused β subunits form tetramers (β_4, also called hemoglobin H, HbH), and the condition is often called HbH disease (suggesting this is a β gene disorder, which it is not). At birth when γ subunits are still being made, excess γ subunits also form tetramers (γ_4, called Hb Bart's). Both HbH and Hb Bart's bind oxygen very tightly and are close to useless for oxygen delivery; and HbH damages red cells and contributes to the anemia. If *four* α genes are missing ($--/--$), virtually all the hemoglobin is Hb Bart's and HbH, and the condition is the "Bart's hydrops fetalis syndrome" (α-thalassemia major). In the past, the fetus usually did not survive to be born, or died shortly after birth. A fetus fortunate enough to have persistent ζ gene expression had a better chance surviving to delivery.

Extremely important is the difference between the two possible configurations of missing two α genes. The arrangement can be either ($-\alpha/-\alpha$) or ($--/\alpha\alpha$). The former, called a "trans" deletion ("across" the alleles inherited from each respective parent), prevails in Africa and the Middle East; the child of such a parent would inherit ($-\alpha$). The other parent, if from the same region, would likely bequeath ($-\alpha$) or ($\alpha\alpha$). The child would then be either ($-\alpha/-\alpha$) or ($-\alpha/\alpha\alpha$), and the condition would be at most α-thalassemia minor.

The prospects are entirely different in Southeast Asia where large α gene deletions take out both α genes on one allele ($--/\alpha\alpha$), referred to as a "cis" deletion ("same side"). The "cis" arrangement is widespread in Cambodia, Thailand, Vietnam, and the Malay Peninsula down to Singapore, and less frequent in the Philippines. A specific deletion, called SEA for Southeast Asia, is the commonest explanation. Smaller deletions taking out just one α-gene ($-\alpha/\alpha\alpha$) also occur in the same region. And additionally, quite common is the non-deletional mutation Constant Spring (α^{CS}), an abnormal lengthening the A2 gene. (α^{CS} genes transcribe poorly, they are functionally impaired, and as the A2 gene directs 60% of α-globin production, α^{CS} is tantamount to an α deletion.)

Thus in Southeast Asia, parents may impart to a child either normal α genes or one of a host of abnormal ones. The child who inherits *one* abnormality from *each* parent may be ($-\alpha/\alpha^{CS}\alpha$), ($\alpha^{CS}\alpha/\alpha^{CS}\alpha$), or ($-\alpha/-\alpha$) with thalassemia minor; or ($--/-\alpha$) or ($--/\alpha^{CS}\alpha$) with HbH disease (α-thalassemia intermedia). And the fetus with ($--/--$) has the Bart's hydrops fetalis syndrome (α-thalassemia major).

The treatment of children and adults with the α-thalassemias is determined individually. Patients with mild to moderate anemia that produces few or no symptoms require no intervention, except possibly advice to avoid iron supplements since iron absorption from the diet is already enhanced in the thalassemias. Those who need repeated transfusions of red cells for HbH disease also need chelators to rid their bodies of excess iron. Family counseling includes specialized blood tests on couples considering pregnancy if routine blood counts suggest thalassemia.

The Bart's hydrops fetalis syndrome requires management in a specialized medical center. If both parents have a cis α deletion ($--$), there is a 25% chance their fetus will have the syndrome. Diagnosis is possible early in pregnancy by week 10–12 by sampling fetal cells in amniotic fluid or placental biopsy. Intrauterine fetal blood transfusion may preserve the fetus during gestation, but early delivery is often required with the attendant problems of premature birth. Then bone marrow stem cell transplantation must follow for a permanent solution. Management is demanding from the standpoints of both medical care and guiding and counseling the parents. Gene therapy is a rapidly emerging technology and likely to be applied in the future. Methods to perpetuate ζ gene expression are of interest. But treatment will not change the baby's germ cell line, which will remain (– –/– –), meaning a child of the next generation will surely inherit one (– –).

The beta thalassemias. Thomas B. Cooley, an early twentieth-century American pediatrician and hematologist practicing in Michigan, was the first to report what came to be called β-thalassemia major. In 1925, he described five affected children. That report was unusual in that this first account came from the United States rather than from the Mediterranean or Middle East where cases were far more prevalent. Clinical features had been described before but not put together into a syndrome. Cooley, relatively unknown beyond his own community, described the marked anemia, abnormal red cells, enlarged spleen, broadened cranial and facial bones, distorted long bones, and jaundice. (Bone deformities result from massive expansion of bone marrow red cell production.) That three of the children died attested to the condition's severity. Cooley did not give a name to this syndrome; others called it "Cooley's anemia", a name that endures today. A profusion of reports followed. For many years a specialized "Cooley's Clinic" operated at the Hadassah Hospital in Jerusalem where I spent several months as a fourth-year medical student in 1972; the clinic morphed into a general thalassemia program.

The β-thalassemias are found in the same geographical distribution as the α-thalassemias, though they appear less frequently in Africa and more commonly in the Mediterranean (Italy, Greece, Cyprus) and the Middle East, extending eastward. They differ fundamentally from the α-thalassemias in important ways. Whereas most α mutations are deletions with no functional α subunit production, among some 350 different β mutations are many single nucleotide substitutions that vary to the extent they impair the production of subunits. The milder ones, with some preserved β subunit production, are referred to as β^+. Those that result in no β subunit production are β^0.

Because one β gene is inherited from each parent, the possibilities are β/β (normal), β^+/β or β^0/β (red cells small, borderline or mild anemia, β-thalassemia minor), β^+/β^+ (spleen may be enlarged, may require transfusion, β-thalassemia intermedia), β^+/β^0 (transfusion dependence common, possible bone disease, β-thalassemia intermedia or major), and β^0/β^0 (uniformly transfusion dependent with bone disease, β-thalassemia major).

Even a fetus with β^0/β^0 will not develop hydrops fetalis because HbF, not built from β subunits, predominates during gestation and suffices for a few months after birth. But then problems for the infant unfold when the baby tries to switch from fetal hemoglobin to adult hemoglobin that it cannot produce. Not only does significant anemia develop, but unused α subunits form tetramers (α_4) that are toxic to developing and mature red cells and shorten their survival. Newborns whose genetics happen to program significant, persistent amounts of γ and δ subunits to make HbF and HbA2 will benefit, as these substitute up to a point for missing HbA and sop up free α subunits; but the infants are still anemic. Even in β-thalassemia minor small excesses of HbF and HbA2 are nearly constant features; their presence is a valuable clue to this thalassemia type as these increases are not part of the α-thalassemia spectrum.

Other β-globin gene region thalassemic disorders. The β-globin gene cluster comprises the consecutive embryonic ε gene, pair of fetal γ (^{G}γ and ^{A}γ) genes, and adult δ and β genes. By far the commonest β-thalassemias involve only the β gene as just described.

A large deletion, however, may also take out the nearby δ gene, producing delta-beta (δβ) thalassemia. Heterozygous individuals, by missing one δ gene, will not have the elevated HbA2 that is so valuable a clue in the diagnosis of β-thalassemia, creating a diagnostic dilemma for the hematologist because routine test results look more like α-thalassemia. Homozygous δβ-thalassemia is exceedingly rare. The individual's parents are often related and carry the same mutation. The affected person may still have only thalassemia minor or intermedia because γ subunit synthesis for HbF production will continue into adulthood. But a homozygous woman's carrying a pregnancy is problematic because her hemoglobin is 100% HbF, and the expected downhill gradient transferring oxygen from mother to fetus will be absent. As a result, fetal growth retardation may be expected, with a risk of other complications. Thus she would best be advised to have a surrogate (whose hemoglobin is HbA) carry the pregnancy.

Larger β region deletions may also take out one or both γ genes, and possibly even the ε gene. All of these in the heterozygous state produce only thalassemia minor, while homozygous εγδβ- and γδβ-thalassemia are presumably incompatible with life.

The β thalassemias are managed in a way similar to the α-thalassemias including red cell transfusion, iron chelation, and family counseling. In thalassemia major, transfusions started early in life can suppress red cell production and thus prevent bone deformities. The recently introduced drug luspatercept can help red cells mature. Parents of infants born with thalassemia major have usually been advised to have their babies undergo bone marrow stem cell transplantation early in life.

Until recently, there had been only one curative treatment for older children and adults with transfusion-dependent β-thalassemia, and that was allogeneic bone marrow stem cell transplantation. It involves identifying a suitably matched donor, drug stimulation to cause stem cells to enter the blood, harvesting the donor's stem cells, obliterating the patient's own marrow with drugs, then intravenous infusion of the donor stem cells. Patients commonly require short-term transfusions and antibiotics, drugs to suppress the immune system rendering the individual vulnerable to infections. If the donor cells attack normal tissues, the resulting "graft-versus-host disease" can be serious and requires management. And the reverse is possible—the patient's immune system may reject the donor stem cells. The procedure is consequently a complex and risky undertaking, and also expensive. It has been most valuable when performed on a young individual with the right mindset (and parental consent in the case of a minor) before serious associated problems have developed. But sometimes a suitably matched donor cannot be found.

Fortunately, gene therapies and gene editing are under intense investigation and are replacing allogeneic stem cell transplantation. The obliteration of the patient's own bone marrow with drugs is similar, but instead of harvesting and then infusing donor stem cells, the patient's own stem cells are harvested, treated ex vivo with the gene therapy, and then infused. Transfusions and antibiotics are often needed for a period of time, but immunosuppression is unnecessary, and graft-versus-host disease will not occur. Questions will remain, however, whether gene editing will be specific only for the intended gene target.

Locatelli and colleagues described 22 patients with transfusion-dependent β-thalassemia variants whose bone marrow stem cells were collected and then, in the laboratory, a β gene programming for $\beta^{87Thr \rightarrow Gln}$ (betibeglogene autotemcel, Zynteglo, Bluebird Bio) was inserted. After the patients' own marrow was obliterated with drugs, the modified stem cells were given back to repopulate the bone marrow. Twenty of the 22 became transfusion independent. This gene therapy was approved by the FDA for β-thalassemia on August 17, 2022.

Several genes, particularly *BCL11A*, are involved in suppressing γ globin chain production starting late in gestation. Exagamglogene autotemcel (Casgevy, Vertex Pharmaceuticals), through CRISPR-Cas9 gene editing, inactivates *BCL11A* and thereby allows resumption of γ globin synthesis, and thus HbF (fetal hemoglobin), compensating for the deficit in HbA production. Locatelli and colleagues reported on 35 patients with transfusion-dependent β-thalassemia followed 12 months or more. The techniques and treatment issues were similar to betibeglogene autotemcel. Among the 35 patients, 32 became transfusion-independent, and the other 3 are still improving. Among the 32, mean hemoglobin was normal at 13.1 g/dL, mean HbF was 11.9 g/dL (90.8% of the total), and 78.0% of marrow stem cells were edited. Off-targeting of the gene therapy seems so far not to be an issue. Casgevy was approved by the FDA on January 16, 2024. Just as a very high level of HbF on an inherited basis may mean impaired maternal-fetal oxygen transfer during pregnancy, so might the high HbF after Casgevy have implications for a woman who wishes to carry a pregnancy. For such a patient, gene therapy with Zynteglo may be preferable.

Hardouin and colleages have described another technique with promise for β-thalassemia patients who have the common mutation IVS1-110 that causes severe disease. A specific "gene editor" can switch the single abnormal adenine (A) nucleotide back to guanine (G), thereby restoring normal β gene function.

The Hemoglobin Variants (E, C, D, O-Arab, and S)

Hemoglobin variants are *intrinsic abnormalities* of the β subunit; they are thoroughly distinct from unbalanced subunit production. In hemoglobins E, C, D, O-Arab, and S, a single amino acid has been switched in the β subunit, producing effects that depend on which amino acid has been changed to what. "Hemoglobinopathy" is a commonly used medical term that means "hemoglobin disease", but a better term is "hemoglobin variant" since these mutations have persisted with a beneficial purpose despite the negatives. The value judgment of "disease" would thus best be left out, but the language is admittedly too well entrenched to do away with easily; and the urge to use it, particularly in the case of sickle cell disease, is understandable.

Hemoglobin E. Hemoglobin E (HbE) is the most prevalent aberrant hemoglobin in the world alongside sickle hemoglobin (HbS). HbE is widespread in northeast India, Sri Lanka, Bangladesh, and throughout Southeast Asia, where in some areas the gene frequency ranges from 30% to 70%. It is the result of a single DNA nucleotide switch resulting in the amino acid substitution of lysine for glutamic acid in the β subunit at position 26. β^E and $\beta^{26Glu \rightarrow Lys}$ are common notations. β^E is a functional subunit, but the abnormal gene programs for an unstable messenger RNA en route to subunit synthesis, thus the amount of β^E produced is low. Furthermore, β^E does not bind as strongly as native β subunits to α subunits and is susceptible to oxidant injury. Heterozygous HbE is called HbAE ($\alpha_2\beta^E\beta$) because hemoglobins are a mixture of mostly HbA and about 25–30% HbE.

Clinically, HbAE is barely noticeable. Red cells are smallish without anemia, but there is some malaria protection. Homozygous HbEE ($\alpha_2\beta^E_2$), often called "hemoglobin E disease" (which markedly overstates its severity), produces still smaller red cells and mild anemia, similar to α- or β-thalassemia minor. Because clinical problems even from HbEE are minimal, this hemoglobin abnormality is a particularly successful antimalarial evolutionary strategy. HbE has proved useful enough that the initiating mutation has occurred more than once in human history; in Southeast Asia its first appearance was roughly 4500 years ago.

HbE-β-thalassemia ($\alpha_2\beta^0\beta^E$ or $\alpha_2\beta^+\beta^E$), by contrast, produces a clinically significant syndrome of anemia, very small red cells, spleen and liver enlargement, and sometimes bone disease (thalassemia intermedia or major). The reason is that not only does the mutated β^E gene transcribe suboptimally, resulting in reduced production of β^E subunits, but the other β gene results in little or no β subunit production at all.

Since α-thalassemias also occur together with HbE, particularly in Southeast Asia, individuals with various combinations of mutations are regular visitors to hematology clinics. Individuals who have HbH disease along with heterozygous or homozygous HbE have HbAE Bart's or HbEF Bart's disease. Affected individuals will be markedly anemic with extremely small red cells and enlarged spleens, and they will often depend on transfusion (thalassemia intermedia or major). These conditions are managed like other significant thalassemic syndromes.

It is difficult to overstate the commonness and significance of the thalassemia disorders in Southeast Asia. For example, about 30–40% of the Thai population carries thalassemia and/or HbE genes; the frequency of α, β, and β^E mutations are respectively 20–30%, 3–9%, and 13% (variable between geographical regions and ethnic groups). In 2022, the population of Thailand was about 66.2 million, and almost half a million, or about 1 in 150 individuals, had thalassemia intermedia or major.

Hemoglobin C. This common hemoglobin variant is as highly adaptive as HbE and results from a glutamic acid to lysine switch at position 6 in the β subunit (β^C, $\beta^{6Glu \rightarrow Lys}$). The heterozygous state (HbAC, $\alpha_2\beta^C\beta$) produces no symptoms and is found in about 17–28% of West Africans, particularly in Burkina Faso and adjacent southern Mali and northern Ghana; the mutation is present in about 2–3% of African-Americans. The homozygous state (HbCC, $\alpha_2\beta^C_2$), "hemoglobin C disease" (which overstates its severity), results in a degree of red cell dehydration; blood smears may reveal hemoglobin crystals and target cells. The hemoglobin-oxygen dissociation curve tends to be right-shifted. Red cell survival is shortened to about 30–35 days (compared to the norm of 120 days), which contributes to the mild anemia. The spleen may be enlarged. Despite these findings, the long-term outlook is good, and pregnancies are usually uncomplicated.

Hemoglobin D and O-Arab. These uncommon hemoglobin variants both result from β subunit substitutions of the glutamic acid at position 121—glutamine in the case of hemoglobin D (HbD) and lysine in hemoglobin O-Arab (HbO-Arab). In both, the heterozygous state is asymptomatic, and the homozygous state is associated with mild anemia. HbAD ($\alpha_2\beta^D\beta$) is found in about 3% of the population of northwest India. HbO-Arab seems to be rare and is found in northern Africa, Sudan, Saudi Arabia, and southeastern Europe.

Hemoglobin S. Hemoglobin S (HbS, β^S) is sickle hemoglobin. It is by far the most consequential and important of all the malaria-adaptive hemoglobin and red cell variants. It commonly occurs in conjunction with other hemoglobin variants and thalassemias. It deserves a chapter of its own, the next one.

Further Reading

[Anon.] Fetal hemoglobin. https://en.wikipedia.org/wiki/Fetal_hemoglobin.
[Anon.] Human genetic resistance to malaria. https://en.wikipedia.org/wiki/Human_genetic_resistance_to_malaria.
[Anon.] Meiosis. https://en.wikipedia.org/wiki/Meiosis#Meiosis.
[Anon.] Rosalind Franklin. https://en.wikipedia.org/wiki/Rosalind_Franklin.
[Anon.] The MNS blood groups. Blood Groups and Red Cell Antigens [Internet]. https://www.ncbi.nlm.nih.gov/books/NBK2274/.
Ayi K, Min-Oo G, Serghides L, et al. Pyruvate kinase deficiency and malaria. N Engl J Med. 2008;358:1805–10. https://doi.org/10.1056/NEJMoa072464.
Badat M, Davies JOJ, Fisher CA, et al. A remarkable case of HbH disease illustrates the relative contributions of the α-globin enhancers to gene expression. Blood. 2021;137:572–5. https://doi.org/10.1182/blood.2021011963.
Baro B, Kim CY, Lin C, et al. *Plasmodium falciparum* exploits CD44 as a coreceptor for erythrocyte invasion. Blood. 2023;143:2016–28. https://doi.org/10.1182/blood.2023020831.
Benz EJ. Hemoglobin variants associated with hemolytic anemia, altered oxygen affinity, and methemoglobinemias. In: Hoffman R, et al., editors. Hematology: basic principles and practice. 3rd ed. New York: Churchill Livingstone; 2000. p. 554–61.
Beutler E. Glucose-6-phosphate dehydrogenase deficiency. N Engl J Med. 1991;324:169–74. https://doi.org/10.1056/NEJM199101173240306.

Chotivanich K, Udomsangpetch R, Pattanapanyasat K, et al. Hemoglobin E: a balanced poly-morphism protective against high parasitemias and thus severe P falciparum malaria. Blood. 2002;100:1172–6. https://doi.org/10.1182/blood.V100.4.1172.h81602001172_1172_1176.

Chui DHK, Fucharoen S, Chan V. Hemoglobin H disease: not necessarily a benign disorder. Blood. 2003;101:791–800. https://doi.org/10.1182/blood-2002-07-1975.

Cooley TB, Lee B. A series of cases of splenomegaly in children with anemia and peculiar bone changes. Trans Am Pediatr Soc. 1925;37:29.

Dawkins R. The selfish gene. New York: Oxford University Press; 1976.

Gong Y, Zhang X, Zhang Q, et al. A natural *DNMT1* mutation elevates the fetal hemoglobin level via epigenetic derepression of the α-globin gene in β-thalassemia. Blood. 2021;137:1652–7. https://doi.org/10.1182/blood.2020006425.

Gregory GL, Wienert B, Schwab M, et al. Investigating zeta globin gene expression to develop a potential therapy for alpha thalassemia major. Blood. 2020;136(Suppl. 1):3–4. https://doi.org/10.1182/blood-2020-142922.

Ingram VM, Stretton AO. Genetic basis of the thalassaemia diseases. Nature. 1959;184:1903–9. https://doi.org/10.1038/1841903a0.

Jiang F, Li D-Z. Outcome of survivors with hemoglobin Bart's hydrops fetalis syndrome: the most severe form of α-thalassemia. Pediatr Transplant. 2021;25:e14090. https://doi.org/10.1111/petr.14090.

Kazazian HH, Waber PG, Boehm CD, et al. Hemoglobin E in Europeans: further evidence for multiple origins of the β^E-globin gene. Am J Human Genet. 1984;36:212–7.

Keith J, Christakopoulos GE, Fernandez AG, et al. Loss of miR-144/451 alleviates β-thalassemia by stimulating ULK1-mediated autophagy of free α-globin. Blood. 2023;142:918–32. https://doi.org/10.1182/blood.2022017265.

Locatelli F, Thompson AA, Kwiatkowski JL, et al. Betibeglogene autotemcel gene therapy for non-β^0/β^0 genotype β-thalassemia. N Engl J Med. 2022;386:415–27. https://doi.org/10.1056/NEJMoa2113206.

Locatelli F, Lang P, Wall D, Meisel R, Corbacioglu S, Li AM, et al. Exagamglogene auto-temcel for transfusion-dependent β-thalassemia. N Engl J Med. 2024;390:1663–76. https://doi.org/10.1056/10.1056/NEJMoa2309673.

Luzzatto L, Mwashungi A, Notaro R. Glucose-6-phosphate dehydrogenase deficiency. Blood. 2020;136:1225–40.

Luzzatto L, Arese P. Favism and glucose-6-phosphate dehydrogenase deficiency. N Engl J Med. 2018;378:60–71. https://doi.org/10.1056/NEJMra1708111.

Miller LH, Mason SJ, Clyde DF, McGinniss MH. The resistance factor to Plasmodium vivax in blacks. The Duffy-blood-group genotype, FyFy. N Engl J Med. 1976;295:302–4. https://doi.org/10.1056/NEJM197608052950602.

Mockenhaupt FP, Ehrhardt S, Cramer JP, et al. Hemoglobin C and resistance to severe malaria in Ghanaian children. J Inf Dis. 2004;190:1006–9. https://doi.org/10.1086/422847.

Murji A, Sobel ML, Hasan L, et al. Pregnancy outcomes in women with elevated levels of fetal hemoglobin. J Matern Fetal Neonatal Med. 2012;25:125–9. https://doi.org/10.3109/14767058.2011.564241.

Ohashi J, Naka I, Patarapotikul J, et al. Extended linkage disequilibrium surrounding the hemoglo-bin E variant due to malarial selection. Am J Human Genet. 2004;74:1198–208.

Paiboonsukwong K, Jopang Y, Winichagoon P, Fucharoen S. Thalassemia in Thailand. Hemoglobin. 2022;46:53–7. https://doi.org/10.1080/03630269.2022.2025824.

Pray LA. DNA replication and causes of mutation. Nat Educ. 2008;1:214.

Prchal JT, Gregg XT. Red cell enzymopathies. In: Hoffman R, et al., editors. Hematology: basic principles and practice. 3rd ed. New York: Churchill Livingstone; 2000. p. 561–76.

Psatha N, Georgakopoulou A, Li C, et al. Enhanced HbF reactivation by multiplex mutagen-esis of thalassemic CD34+ cells in vitro and in vivo. Blood. 2021;138:1540–53. https://doi.org/10.1182/blood.2020010020.

Randhawa ZI, Jones RT, Lie-Injo LE. Human hemoglobin Portland II (ζ2β2): isolation and characterization of Portland hemoglobin components and their constituent globin chains. J Biol Chem. 1984;259:7325–30. https://doi.org/10.1016S0021-9258(17)39875-7

Sae-ung N, Goonnapa F, Kanokwan S, Supan F. Alpha(o)-thalassemia and related disorders in Northeast Thailand: a molecular and hematological characterization. Acta Haematol. 2007;117:78–82. https://doi.org/10.1159/000096857.

Taher AT, Musallam KM, Cappellini MD. β-Thalassemias. N Engl J Med. 2021;384:727–43. https://doi.org/10.1056/NEJMra2021838.

Traivaree C, Boonyawat B, Monsereenusorn C, Rujkijyanont P, Photia A. Clinical and molecular genetic features of Hb H and AE Bart's diseases in central Thai children. Appl Clin Genet. 2018;11:23–30. https://doi.org/10.2147/TACG.S161152.

Wambua S, Mwangi TW, Kortok M, et al. The effect of α+-thalassaemia on the incidence of malaria and other diseases in children living on the coast of Kenya. PLoS Med. 2006;3:e158. https://doi.org/10.1371/journal.pmed.0030158.

Watson JA, Leopold SJ, Simpson JA, Day NPJ, Dondorp AM, White NJ. Collider bias and the apparent protective effect of glucose-6-phosphate dehydrogenase deficiency on cerebral malaria. elife. 2019; https://doi.org/10.7554/eLife.43154.

Watson JD, Crick FHC. Molecular structure of nucleic acids: a structure for deoxyribose nucleic acid. Nature. 1953;171:737–8. https://doi.org/10.1038/171737a0.

Weatherall DJ. Toward an understanding of the molecular biology of some common inherited anemias: the story of thalassemia. In: Wintrobe MM, editor. Blood pure and eloquent: a story of discovery, of people, and of ideas. New York: McGraw-Hill; 1980. p. 373–414.

Weatherall DJ. The thalassemias. In: Beutler E, et al., editors. Williams Hematology. 6th ed. New York: McGraw-Hill; 2001. p. 547–80.

Yen A, Zappala Z, Altshuler D. Specificity of CRISPR-Cas9 editing in exagamglogene autotemcel. N Engl J Med. 2024;390:1723–25.

Chapter 8
Sickle Hemoglobin and Sickle Cell Disease

On the morning of May 5, 1910, the 49-year-old Dr. James B. Herrick of Chicago was attending the 25th annual meeting of the Association of American Physicians at the renowned Willard Hotel in Washington, D.C., when he presented the unusual case of an anemic patient whose blood had numerous "sickle-shaped" red cells. It was the first description of sickle cell disease. Those in attendance were numerous prominent internists who gave Dr. Herrick's presentation practically no attention.[1]

Dr. Herrick published the case in extraordinary detail the same year in the *Archives of Internal Medicine*, with superb photomicrographs of the red blood cells. He described a 20-year-old Black male from Grenada in the West Indies whom he had followed as a patient for about 6 years with "unusual blood findings, no duplicate of which I have ever seen described." He speculated that the condition was "some peculiar physical or chemical condition of the blood." His patient had many of the cardinal features of what was later established as sickle cell disease, and Herrick, by describing the shapes of the red cells, unwittingly gave the condition an identifier that soon stuck as a name for the disorder. He detailed the young man's symptoms: anemia with a hemoglobin level only half normal, jaundice, leg ulcers, but no enlarged spleen, and a blood smear with a "large number of thin, elongated, sickle-shaped and crescent-shaped forms".

Tellingly, he described a classic episode of what is now referred to as a "sickle pain crisis". His patient had back pain, aching in his legs and arms, severe pain in the upper abdomen, vomiting, worsening jaundice, and mild fever, all of which left him hospitalized for 2 months. Pain crises are now known as just one of the several kinds of dreadful vascular complications that afflict individuals with sickle cell disease and which set the condition apart from all other disorders of hemoglobin and red cells that evolved to counter malaria. It is impossible to overstate the misery and the tragedy of pain, debility, and premature death that sickle cell disease has

---

[1] As recounted in 1980 by C. Lockard Conley, then head of the Hematology Division at the Johns Hopkins University School of Medicine, Baltimore.

© The Author(s), under exclusive license to Springer Nature
Switzerland AG 2024
M. H. Rosove, *Life's Blood*, https://doi.org/10.1007/978-3-031-61150-6_8

Fig. 8.1 James B. Herrick and his photomicrographs (1910) (Credit: Studio portrait, Moffett Studio, Chicago, ca. 1915 [left], and Herrick JB. *Archives of Internal Medicine* 1910;6:517–521 [right])

wrought, as well as the emotional turmoil it has brought to the loved ones on whom they have depended. A current estimate is that every year about 300,000 babies are born worldwide with sickle cell disease, 75% of them in sub-Saharan Africa, the rest widely distributed (Fig. 8.1).

Advances in knowledge after 1910 came in dribs and drabs over the next several decades. Doctors frequently reported seeing sickle cells, but surprisingly, the first case of sickle cell disease in Africa was not described until 1932. Soon, family studies of the parents of affected individuals found that in nearly all cases the parents had red cells that could be made to sickle in the laboratory, even though they had little or nothing in the way of symptoms or signs of the disease. They had what came to be known as sickle trait. The frequency of sickle trait in Africa was found to be very high: 12.9% in Rhodesia (now Zimbabwe), 20% in The Gambia, Gold Coast (now Ghana), Nigeria, and Cameroons, and 25% in Belgian Congo (now Congo). In some local tribes, prevalence was as high as 40%. (The rate in Blacks in the United States is about 8%.) The condition conformed to typical Mendelian inheritance where those with sickle trait carry one copy of the sickle gene, and those with sickle cell disease typically have two.

If one in five of a population were to have one sickle gene, then one in 25 couples would carry one gene each, and an unlucky child would have a one in two chance of inheriting the gene from either parent, and thus a one in four chance of getting it from both and developing the disease. In that scenario, one in every 100 births would yield a baby with sickle cell disease. And yet Africans with sickle cell disease were rarely described even through the 1960s. A number of explanations were put forth; but the true answer ultimately lay in discovering precisely the causes of death in young children where falciparum malaria and other endemic and sickle cell-associated infections were common and created diagnostic confusion. Sickle cell disease was not rare. It simply killed so many children early in life that many cases were not detected, particularly by physicians not routinely caring for pediatric patients. Once more careful diagnoses were made, doctors determined that half of all children in Congo with the disease reported in 1955 had died by the age of 5. In Zambia, even as late as 1970, half were gone by age 3. In Ghana, where a specialty clinic in 1971 emphasized living standards, nutrition, treatment of infections, and health maintenance, half the children lived to 10 years or older, a modest but not-to-be-disregarded improvement.

In the early years of research and discovery, one of the unsettled questions was why the sickling phenomenon was so prevalent. In 1946, E. A. Beet in southern Rhodesia (now Zimbabwe) made an important, new observation—only 10 of 102 (9.8%) with the sickling phenomenon (sickle trait) developed malaria, whereas 75 of 491 (15.3%) without it did. The difference was statistically significant. By the mid-1950s, as evidence accumulated, the answer progressively came into hand. From the same region, P. Brain suggested that "red cells in sicklers offer a less favourable environment for malarial parasites." A. C. Allison found that among 290 young children in malaria areas surrounding Kampala, Uganda, only 12 of 43 (27.9%) with sickling carried malarial parasites, while 113 of 247 (45.7%) without sickling did. Allison further observed that at Lake Victoria in Kenya, where malarial rates were especially high, there was a correspondingly higher-than-expected rate of sickle trait. And J. P. Mackey and F. Vivarelli in Dar es Salaam, Tanganyika (now Tanzania), speculated that carrying one sickle cell mutation was advantageous for survival because it reduced susceptibility to falciparum malaria. The protective value against malaria of inheriting one copy of the sickle gene was increasingly appreciated, and what was previously considered only an idea was now broadly accepted as fact, which it remains today. Individuals born with sickle cell disease were the unfortunate price the larger population would pay. No other hemoglobin variant has exacted such a price.

A series of crucial observations about sickle hemoglobin were made from the 1920s to 1970s. In 1927, E. Vernon Hahn and Elizabeth Biermann Gillespie at the University of Indiana, Indianapolis, reported a case of sickle cell disease with astute laboratory examinations concerning the role of oxygen in the sickling phenomenon. They observed that not all the red cells were sickled and thus surmised that sickle cells were not made that way, but rather external factors must be operative. They tested cold, heat, light, and the milieu of the blood in various ways, none of which were involved. But they noticed that in a vertical tube of blood, the red cells settling out most by gravity showed more sickling, and they conjectured that oxygen deprivation in the deep recesses of the tube might be operative. And they found that the sickled cells could be restored to their normal shape by "pure oxygen". They further noted that sickling occurred when the partial pressure of oxygen dropped below

45 mmHg, a threshold subsequently confirmed by others.[2] And that red cell "ghosts" (cells from which hemoglobin was removed) did not sickle. These experiments established that hemoglobin was the culprit in sickle cell disease and that sickling depended on oxygen tension, observations that would have lasting, enormous impact on future research and patient care. Hahn was a young surgeon recently out from training, and Gillespie was a medical intern: their work was quite an accomplishment for a couple of young physicians. Subsequent work by others showed that oxygen tension had to be much lower to induce sickling in sickle trait.

In 1933, Lemuel W. Diggs of Memphis, whose research focused on sickle cell disease, presented his findings from autopsied patients and described with great insight the essential pathophysiological cause of the vascular complications. He wrote, "A possible explanation of the capillary engorgement is that the elongated and spiked cells interlock and pass with more difficulty through narrow spaces than do normal cells.... Since the distortion is increased under conditions of anoxemia [low oxygen tension], it is reasonable to assume that it will be greatest in tissue where there is stasis [slowing of blood flow].... The sudden pains experienced by patients with sickle cell anemia, which often disappear as mysteriously as they come, may in part be explained by this capillary blockade."

The sickling phenomenon in the smallest blood vessels of the body came to be understood as the root cause of various complications setting sickle cell disease apart from the thalassemias and other miscellaneous disorders of premature red cell destruction. Sickling is a vicious, cascading cycle feeding on itself of small blood vessel obstruction, sluggish blood flow, worsening local oxygen tension, and then more sickling. In the youngest children, sickling in the spleen, where the flow of blood out of the organ is particularly likely to be blocked, may suddenly cause it to engorge with blood, leading to abrupt worsening of the anemia with fall in blood volume that can lead to shock, an emergency called "splenic sequestration crisis". In older children and adults, repeated occlusive vascular insults to the spleen cause it to transform into non-functional scar tissue, explaining the predilection to certain serious life-threatening infections, and also the common but otherwise inconsequential finding of circulating nucleated red cells that are normally not seen. Vascular occlusions are responsible for pain crises, painful leg ulceration, blood engorgement of the penis, venous blood clotting, stroke, and injury to the retinas, heart, kidneys, and bones.

A complication especially worthy of note is the "acute chest syndrome". Patients with the syndrome have cough, shortness of breath, chest X-ray abnormalities, falling hemoglobin-oxygen saturation, and a cascade of sickling with dysfunction in multiple organ systems. The syndrome, which is a medical emergency, has a high mortality rate if not properly treated. Its commonest triggers are infections and

---

[2] The mean oxygen tension in blood returning from the tissues is lower, about 37 mm Hg, meaning that in the small-caliber capillaries where oxygen-carbon dioxide exchange occurs the oxygen tension is in a range threatening sickling. Circulation times are usually fast enough, however, that most of the time hemoglobin reaches the lungs before sickling gets underway.

simultaneous pain crises, particularly when bone and bone marrow are involved and fat and marrow tissue enter the circulation and lodge in the lungs.[3]

In the late 1940s, when researchers had yet to understand the nature of sickle hemoglobin, Linus Pauling at Caltech became interested in the question, and three of his postdoctoral students, Harvey A. Itano, S. J. Singer, and Ibert C. Wells, employing a technique already being used to separate proteins of differing electrical charge, showed that sickle cell disease and normal adult hemoglobin were different. Sickle trait was a mixture of both. The work was presented in 1949 to the American Society of Biological Chemists and to the National Academy of Sciences and published the same year in *Science*. The four authors had thus wittingly established sickle cell disease as the first "molecular disease", that is, one with a definitive molecular-clinical correlation, and all this before Max Perutz had elucidated the structure of hemoglobin. The Caltech group's discovery opened a broad new field of medical research into conditions suspected of having an inherited basis.

What caused the difference in electrical charge? The reason was found by Vernon M. Ingram in Cambridge in 1956 where he had been recruited by Perutz as a protein chemist to assist Perutz and Kendrew with their respective crystallographic studies of hemoglobin and myoglobin.[4] Ingram's success with the analysis of sickle hemoglobin came with a combination of techniques whereby he enzymatically digested hemoglobin into short amino acid sequences and analyzed their electrical charge differences and amino acid compositions. Five sickle cell disease patients' hemoglobins all yielded the same result: the difference between sickle and normal adult hemoglobin was just a single amino acid. One glutamic acid was replaced by valine. With others' successful sequencing of amino acids in the α- and β-globin chains, the amino acid switch was established in 1958 at position 6 of the β-globin chain. Sickle hemoglobin is commonly notated as HbS, and the β subunit as β^S or $\beta^{6Glu \to Val}$. At the DNA level, a thymine for adenine switch in the middle of the nucleotide triplet that codes for glutamic acid is responsible. The mutation is now known to have occurred somewhere between 7000 and 22,000 years ago in the rainforest of present day Cameroon. Slave trades to the Middle East and southern Asia began in the seventh century CE and to the Americas in the fifteenth century, distributing the sickle mutation far and wide.

Identifying the change in hemoglobin's primary structure defining sickle hemoglobin was an astounding achievement, but it could not in and of itself explain the physical property of sickling. After all, numerous other single amino acid substitutions in the thalassemias and other hemoglobin disorders did not cause sickling,

---

[3] A different kind of complication unrelated to vascular occlusion is the "aplastic crisis", meaning the bone marrow suddenly ceases to produce the numbers of red cells necessary to keep up with shortened red cell survival. The commonest cause is infection with parvovirus B19, the cause of "fifth disease" in children, which also occurs in adults. Other infections may do the same. Recovery of marrow function is the rule, but during the crisis anemia can become life-threatening.

[4] Ingram was born in 1924 in Breslau; his given name was Werner Immerwahr. As a Jew under the Third Reich he was not allowed to go to school, his family fled to London before the Nazis liquidated Breslau's Jewish community, and in London he anglicized his name.

even hemoglobin C, also a single amino acid substitution at the very same position 6 in the β-globin chain (but lysine instead of valine). Obviously, there had to be something unique about the switch to valine. A clue was the finding by B. C. Wishner and colleagues that among the respective 141 and 146 amino acids in the α and β subunits, the sickling phenomenon involved just four points of inter- and intra-molecular contact, one of them being the valine substitution.

As for sickling itself, deoxygenated sickle hemoglobin forms polymers that twist together into cables that stack, align, and distort the red cell into its sickle shape, making red cells inflexible and blood more viscous. Even though sickling is revers-ible, with repeated sickling and unsickling the red cell membrane becomes dam-aged. All these changes not only make for blood vessel obstruction, they also shorten the life of red cells to an average of just 20 days, one-sixth of normal.

The tendency to sickle is directly proportional to the sickle hemoglobin concen-tration. Perhaps it is an irony that it is harmful only at high concentration and low oxygen tension because it otherwise functions normally in oxygen transport and delivery. To set some benchmarks as a basis for discussion, the sickle hemoglobin concentration in sickle cell disease is about 90% to 95%, and in sickle trait about 35% to 40%. Co-inheritance of another β mutation will position the sickle hemoglo-bin concentration somewhere in between.

A smorgasbord of hemoglobin and red cell mutations exist across the malaria belt. Thus it should come as no surprise that sickle hemoglobin is often co-inherited with other variants. The frequency of any particular combination is related to the occurrence of each in their respective geographical regions. Thus sickle cell disease inherited with "trans" α-thalassemia minor ($-\alpha/-\alpha$) in Africa is common—and it is advantageous because, without enough α subunits to pair up with β^S subunits, the amount of sickle hemoglobin per red cell is reduced. Heterozygous HbS coupled with HbC or β-thalassemia minor are also common in Africa and the Middle East. Others are rarer.

After homozygous HbSS (sickle cell disease), the most serious is HbS-β^0-thalassemia. HbS concentration, as with HbSS, is 90% or greater because, aside from β^S, no other kind of β subunit is produced. HbS-β^+-thalassemia is less serious since some HbA is made. Next serious are HbSD and HbSO-Arab (both rare); they resemble HbSS because these variants engage well with HbS in polymerization. HbSC produces less severe disease.

Some people have "hereditary persistence of hemoglobin F" that reduces HbS concentration by displacing it. HbSE is mild or even asymptomatic because HbE does not promote polymerization; and the HbE level of about 30% keeps HbS down to about 60% to 65%. However, individuals with HbSE or sickle trait (HbAS) can experience sickling complications in the spleen or kidneys when dehydration is combined with reduced arterial partial oxygen tension, as at high altitude, during airline travel, or with lung disease.

The treatment of sickle cell disease is multifaceted and must be tailored to the specific patient and problem at hand. Ever since "hereditary persistence of hemo-globin F" was found to ameliorate the seriousness of sickle cell disease, there has been interest in increasing HbF in red cells. The convenient oral drug hydroxyurea, which rarely causes side effects, was introduced into clinical practice in 1967 and has since been used to treat several bone marrow disorders. When it was found to elevate HbF, small numbers of sickle cell disease patients were tried on it and seemed to benefit. Then in 1995, Samuel Charache and colleagues of the Investigators of the Multicenter Study of Hydroxyurea in Sickle Cell Anemia reported on 299 patients. Half were given the drug, and half placebo. The treated patients had a doubling of red cells containing HbF, roughly a halving of numbers of pain crises and acute chest events, with reduced need for transfusion. (And recently reported was a reduced risk of falciparum malaria during hydroxyurea treatment in sub-Saharan Africa.) The FDA approved the drug in adults in 1998 and extended the approval to pediatric patients in 2017. It is the most widely used supportive medication in sickle cell disease and is a first-line treatment. People who take it commonly achieve HbF levels of 10% to 20% or more. Three other drugs have received FDA approval and several more are under investigation, working by mechanisms other than raising HbF.[5]

Sickle crises are always managed with supplemental oxygen, hydration, pain relief, nutrition, careful surveillance for possible emerging acute chest syndrome, and looking for infection as the trigger. Patients often need red cell transfusion or whole blood exchange, and treatments to prevent blood clots and infections. Preventive and supportive measures may include iron chelation (to get rid of excess iron from previous red cell transfusions), non-iron-containing vitamin supplements (especially folic acid because bone marrow so active producing red cells needs more), vaccinations, adequate hydration during airline flights, genetic counseling, management of depression and psychosocial issues (which are common), and chronic pain control. Some patients require surgical management of gallstones, a common problem whenever red cell survival is chronically shortened. (Short red cell survival increases delivery of bilirubin to the bile from red cell breakdown, beyond bilirubin's solubility.) Some may develop antibodies to red cell antigens on other people's red cells after transfusion or pregnancy, limiting which donated units of red cells can be given, and these individuals need special attention from blood banks. When both prospective parents have sickle trait (or another concerning mutation), in vitro fertilization and embryo selection are possible. Pregnancy requires careful management by maternal-fetal medicine specialists. Day-to-day management is handled by hematologists, primary care physicians, and nurse practitioners, often in dedicated clinics.

Until recently, there had been only one "curative" sickle cell disease treatment, one that could reduce or eliminate sickle crises, and that was allogeneic bone marrow stem cell transplantation, considerations being the same as described in the last

---

[5] Voxelotor increases HbS oxygen affinity and inhibits sickle hemoglobin polymerization and thereby somewhat relieves red cell destruction. L-glutamine is anti-oxidant. Crizanlizumab inhibits sickle cell adhesion to blood vessels. Recent experimental therapies generating interest include mitapivat that depletes 2,3-BPG, thereby increasing HbS oxygen affinity; canakinumab which blocks interleukin-1β, thus lessening the inflammatory response to sickling; and vamifeport that limits iron availability for red cells to make HbS.

chapter for transfusion-dependent β-thalassemia except that allogeneic transplantation is especially risky in sickle cell disease patients because of preexisting organ dysfunction. Donor stem cells amounting to just 20% to 25% of the total may suffice to lower the amount of circulating sickle hemoglobin enough to relieve the clinical consequences of sickling.

Because of the problems inherent in allogeneic stem cell transplantation, genetic therapies have supplanted it and are at the forefront of therapeutics. They are not just the future of definitive care, but now the present definitive care. They do not involve stem cells from another person and hold the promise of lifelong relief. Genetic therapy in sickle cell disease has so far taken one of three approaches similar to β-thalassemia: providing a gene that will make a HbA look-alike; unbridling HbF manufacture that has been suppresed since birth; or editing the abnormal DNA codon programming for the disease-causing valine at position 6 of the β subunit. The general principles of gene therapies in sickle cell disease, including questions of off-targeting, are similar to those in β-thalassemia. But whereas β-thalassemia patients are entirely dependent on the hemoglobin product of the gene therapy, sickle cell disease patients are not as demanding: lowering the HbS concentration from over 90% down to 40% to 50% will relieve most or all new disease manifestations.

The results of genetic therapies to date have been remarkable. Kanter and colleagues inserted into bone marrow stem cells, by means of a harmless viral vector, a modified β gene similar to the one applied in β-thalassemia, also programming for $\beta^{87Thr \rightarrow Gln}$, a HbA look-alike with a normal hemoglobin-oxygen dissociation curve (lovotibeglogene autotemcel, Lyfgenia, Bluebird Bio). In 25 treated, evaluable sickle cell disease patients, the modified hemoglobin constituted 40% of the total, and the patients, who previously had had a median of 3.5 vascular crises per year, now were event-free during intermediate-term follow-up. The FDA approved Lyfgenia for sickle cell disease on December 8, 2023. Further modifications may ultimately enable the engineered β subunits to out-compete β^S for α subunits, further reducing sickle hemoglobin concentration.

Another effective avenue has been restoration of HbF production. Esrick and colleagues inserted messenger RNA with a lentiviral vector to turn off the gene *BCL11A* in 6 sickle cell disease patients. That permitted the restitution of γ subunit synthesis and making of HbF. After treatment, the patients had "stable and robust" HbF production ranging from 20.4% to 41.3% of the total, broadly distributed in red cells, with displacement of HbS.

Frangoul and colleagues took another approach to stepping up HbF production. They used the same exagamglogene autotemcel (Casgevy, Vertex Pharmaceuticals), through CRISPR-Cas9 gene editing, as in β-thalassemia patients, to inactivate *BCL11A*. Forty-four sickle cell disease patients achieved HbF levels over 40%, broadly distributed in 93% of red cells; a mean of 86.1% of bone marrow stem cells were successfully edited. Of 30 patients followed 12 months or more who had had a mean of 3.9 sickle crises per year, 29 became event-free. Pre-treatment red cell destruction abated, and hemoglobin levels normalized. Given that HbF levels after treatment in sickle cell disease patients are much lower than in β-thalassemia patients, concern over oxygen transfer from mother to fetus during pregnancy might be less. The FDA approved Casgevy on December 8, 2023, the same day as it did Lyfgenia. Long-term follow-up studies of both are ongoing.

In a different kind of genetic management, Li and colleagues reported a "vectorized prime editing system" in a mouse model and in human bone marrow stem cells in the laboratory to repair the abnormality in the DNA nucleotide triplet programming position 6 of the β subunit. By switching the abnormal thymine back to adenine, the abnormal valine reverted back to the normal glutamic acid of HbA.

In the distant past very few children lived past the age of 5–10 in sub-Saharan Africa, but the outlook today for children and adults with sickle cell disease everywhere is vastly different. As complex and difficult as management remains, most survive well into adulthood thanks to preventive measures and expert management of crisis situations. Obstacles remain. Patients may be reluctant to keep regular follow-ups with their health care providers during the stretches of time they are well, which result in avoidable complications. Education and medical care, which are of paramount importance, are lacking in various parts of the world. And it will be challenging to get the most promising and advanced genetic technologies to areas of the world where the greatest number of potential beneficiaries reside, especially considering how expensive the treatments are sure to be.

Will a reduction in sickle hemoglobin concentration from treatment raise the risk of serious complications from falciparum malaria? The answer is unclear, but antimalarial strategies would best preemptively be advanced hand-in-hand with sickle cell treatments. While there is no way to overstate how much misery sickle cell disease has caused in the past, it is also impossible to overstate the fact that medical progress has radically changed the outlook for the countless people still affected today, and for newborns in the future.

Further Reading

Abraham AA, Tisdale JF. Gene therapy for sickle cell disease: moving from the bench to the bedside. Blood. 2021;138:932–41. https://doi.org/10.1182/blood.2019003776.

Allison AC. Protection afforded by sickle-cell trait against subtertian malarial infection. Br Med J. 1954;1:290–4. https://doi.org/10.1136/bmj.1.4857.290.

[Anon.] Linus Pauling. https://en.wikipedia.org/wiki/Linus_Pauling.

Barclay GPT, Huntsman RG, Robb A. Population screening of young children for sickle cell anaemia in Zambia. Trans R Soc Trop Med Hyg. 1970;64:733–9.

Beet EA. Sickle cell disease in the Balovale District of Northern Rhodesia. East Afr Med J. 1946;23:75–86.

Bertles JF, Milner PFA. Irreversibly sickled erythrocytes: a consequence of the heterogeneous distribution of hemoglobin types in sickle-cell anemia. J Clin Invest. 1968;47:1731–41.

Beutler E. The sickle cell diseases and related disorders. In: Beutler E, et al., editors. Williams Hematology. 6th ed. New York: McGraw-Hill; 2001. p. 581–605.

Brain P. Sickle-cell anaemia in Africa. Br Med J. 1952;2:880.

Bunn HF. Oxygen delivery in the treatment of anemia. N Engl J Med. 2022;387:2362–5. https://doi.org/10.1056/NEJMra2212266.

Charache S, Terrin ML, Moore RD, et al. Effect of hydroxyurea on the frequency of painful crises in sickle cell anemia. N Engl J Med. 1995;332:1317–22. https://doi.org/10.1056/NEJM199505183322001.

Conley CL. Sickle-cell anemia—the first molecular disease. In: Wintrobe MM, editor. Blood pure and eloquent: a story of discovery, of people, and of ideas. New York: McGraw-Hill Book Company; 1980. p. 318–71.

Diggs LW, Ching RE. Pathology of sickle cell anemia. South Med J. 1934;27:839–45. https://doi.org/10.1097/00007611-193410000-0000.

Esoh K, Wonkam A. Evolutionary history of sickle-cell mutation: implications for global genetic medicine. Hum Mol Genet. 2021;30:R119–28. https://doi.org/10.1093/hmg/ddab004.

Esrick EB, Lehmann LE, Biffi A, et al. Post-transcriptional genetic silencing of BCL11A to treat sickle cell disease. N Engl J Med. 2021;384:205–15. https://doi.org/10.1056/NEJMoa2029392.

[FDA News Release] FDA approves first gene therapies to treat patients with sickle cell disease. 8 Dec 2023.

Frangoul H, Altshuler D, Cappellini Y-S, et al. CRISPR-Cas9 gene editing for sickle cell disease and β-thalassemia. N Engl J Med. 2021;384:252–60. https://doi.org/10.1056/NEJMoa2031054.

Hahn EV, Biermann GE. Sickle cell anemia. Report of a case greatly improved by splenectomy. Experimental study of sickle cell formation. Arch Int Med. 1927;39:233–54. https://doi.org/10.1001/archinte.1927.00130020072006.

Herrick JB. Peculiar elongated and sickle-shaped red blood corpuscles in a case of severe anemia. Arch Int Med. 1910;6:517–21.

Ingram VM. The chemical difference between normal human and sickle cell anaemia haemoglobins. Conference on Hemoglobin. Publication 557. National Academy of Sciences—National Research Council; 1958. p. 233–8.

Ingram VM. Hemoglobin and its abnormalities. Springfield: Charles C. Thomas; 1961.

Kanter J, Walters MC, Krishnamurti L, et al. Biologic and clinical efficacy of LentiGlobin for sickle cell disease. N Engl J Med. 2022;386:607–28. https://doi.org/10.1056/NEJMoa2117175.

Kavanagh PL, Fasipe IA, Wun T. Sickle cell disease: a review. Am Med Assoc J. 2022;328:57–68. https://doi.org/10.1001/jama.2022.10233.

Konotey-Ahulu FID. Computer assisted analysis of data on 1,697 patients attending the Sickle Cell/Haemoglobinopathy Clinic of Korle-Bu Teaching Hospital, Accra, Ghana. Ghana Med J. 1971;10:241–60.

Lambotte-Legrand J, Lambotte-Legrand C. Anémie drépanocytaire et homozygotism (À propos de cas de300 déces). Ann Soc Belg Med Trop. 1955;35:47–51.

Lambotte-Legrand J, Lambotte-Legrand C. Le prognostic de l'anémie drépanocytaire au Congo Belge (À propos de cas de 150 déces). Ann Soc Belg Med Trop. 1955;35:53–7.

Mackey JP, Vivarelli F. Sickle-cell anaemia. Br Med J. 1954;1:276.

Olupot-Olupot P, Tomlinson G, Williams TN, et al. Hydroxyurea treatment is associated with lower malaria incidence in children with sickle cell anemia in sub-Saharan Africa. Blood. 2023;141:1402–10. https://doi.org/10.1182/blood.2022017051.

Pauling L, Itano HA, Singer SJ, Wells IC. Sickle cell anemia, a molecular disease. Science. 1949;110:543–8. https://doi.org/10.1126/science.110.2865.543.

Sharma A, Beolens J-J, Cancio M, et al. CRISPR-Cas9 editing of the HBG1 and HBG2 promoters to treat sickle cell disease. N Engl Med J. 2023;389:820–32. https://doi.org/10.1056/NEJMoa2215643.

Wishner BC, Ward KB, Lattman EE, Love WE. Crystal structure of sickle-cell deoxyhemoglobin at 5 Å resolution. J Mol Biol. 1975;98:179–94.

Zimmerman SA, O'Branski EE, Rosse WF, Ware RE. Hemoglobin S/O(Arab): thirteen new cases and review of the literature. Am J Hematol. 1999;60:279–84. https://doi.org/10.1002/(sici)1096-8652(199904)60:4<279::aid-ajh5>3.0.co;2-2.

Chapter 9
Hemoglobin and Red Cell Adaptation When Oxygen Is Lacking

A number of conditions challenge hemoglobin's ability to acquire and deliver oxygen. One is high altitude, where atmospheric pressure and oxygen tension are reduced, with fewer oxygen molecules in every breath. Serious lung disease can restrict breathing, or impose a barrier between the alveoli and blood, or create a mismatch within the lungs between the places with the best ventilation and those with the best blood flow. Some congenital heart diseases and vascular abnormalities permit some of the blood returning to the heart to bypass the lungs without being reoxygenated. Anemias of diverse causes reduce the blood's oxygen carrying capacity. And some congenital hemoglobin variants do not bind or release oxygen normally. Investigations into the nature of all of these have generated mountains of research papers and have uncovered fascinating details about how humans and other animals respond or adapt to oxygen adversity.

The Challenges of High Altitude Adventure

High altitudes have long been siren calls for the most able-bodied and adventurous.[1] The exploration literature is filled with detailed descriptions of colorful, dramatic exploits in the highest mountains, where unimaginable hardships and challenges have drawn either the courageous trying to achieve something they believed impossible, or the seekers of scientific knowledge, national pride, or fame. These ventures have also spurred meaningful research into human high-altitude adaptation.

European mountaineering gained solid forward momentum in the eighteenth century by new descriptions of Chamonix in southeastern France near Mont Blanc,

---

[1]Altitudes are given in meters. One meter is 3.281 feet. To convert meters to feet, a close approximation is to multiply by 3, then add 10%. Temperatures are given in °C. To convert °C to °F, multiply by 1.8 and add 32.

© The Author(s), under exclusive license to Springer Nature Switzerland AG 2024
M. H. Rosove, *Life's Blood*, https://doi.org/10.1007/978-3-031-61150-6_9

followed by the first ascent of Mont Blanc (4810 m) in 1786, and then the first ascent of the Matterhorn (4478 m) in 1865. With these high peaks conquered, enthusiasts of altitude, especially the French, developed the sport of ballooning, pushing altitudes ever higher, risking the shortness of breath and cognitive impairments that come with oxygen deprivation. In 1875, the balloon *Zénith* reached about 7900 m, but two of the three passengers died of asphyxiation. They had carried oxygen equipment in the basket, but never used it. The renowned French physiologist Paul Bert that same decade provided irrefutable evidence that altitude and symptoms from lack of oxygen were connected.

The twentieth century brought a steady stream of attempts on the world's highest peaks in the Himalayas. Ever since Edmund Hillary and Tenzing Norgay achieved the first summiting of Mount Everest on May 29, 1953, many have sought that alluring, highest point on the globe. As of July 2022, 11,346 ascents to the summit (8848.86 m) had been accomplished by 6098 people at considerable financial expense, long periods of planning and acclimatization to altitude, and significant risk to life itself. In an average year nowadays, about 700 to 800 people make the attempt, and one in 60 die trying. Every problem of this kind of climb is compounded by the fact that a brain operating on limited oxygen has crippled judgment and decision-making ability, so climbers have to navigate somehow without being able to think clearly. Journalist and climber Jon Krakauer brought home for the armchair adventurer the Everest disaster of May 10–11, 1996, in his classic first-hand account *Into Thin Air*, a narrative detailing the fallout of a violent storm on those fateful days. Among the extreme features of the environment, he talked about the debilitating hypoxic atmosphere that could only be slightly relieved by carefully rationed supplemental oxygen. Krakauer summited during that tragic episode and described his mindset while in the formidable "Death Zone", commonly defined as an altitude above 8000 m.[2] "The ratio of misery to pleasure was greater by an order of magnitude than any other mountain I'd been on; I quickly came to understand that climbing Everest was primarily about enduring pain. And in subjecting ourselves to week after week of toil, tedium, and suffering, it struck me that most of us were probably seeking, above all else, something like a state of grace."

Until the 1970s everyone assumed that summiting Everest without supplemental oxygen was impossible. But then some climbers, the South Tyrol mountaineer Reinhold Messner among them, speculated that perfectly fit, experienced, and properly acclimatized—and, of course, lucky—individuals ought to be able to succeed without it. That would spare the need for carrying oxygen bottles, concern over

---

[2] Features of the environment and human responses in the "Death Zone" include the lack of enough oxygen, the difficult terrain, possible need to trek in the dark, complications of clothing and equipment, unexpected icefalls (including earthquake), intense solar radiation, snowblindness, high winds, hypothermia and frostbite, gastrointestinal complaints (vomiting, diarrhea), sleep apnea, pulmonary edema, altered cognition (from cerebral edema or hyperviscosity of blood from markedly elevated red cell numbers), difficulty of preparing food and water with malnourishment and dehydration, icy condensation from exhaled breath, fear, depression, intermixing personalities and cultures of one's companions, and mutual dependence but limited ability to help or be helped in an emergency.

running out of the precious gas, and littering mountainsides with exhausted containers. But the risk would be enormous. On May 8, 1978, Messner and his Austrian climbing partner Peter Habeler succeeded and made worldwide news, some commentators now glibly asserting that in the future a "true" ascent of Everest would require relying only upon the oxygen nature provided. Some even suggested that to carry oxygen was "cheating". On August 20, 1980, Messner achieved the first solo Everest summiting, on the Tibetan north face, without oxygen, and during the adverse monsoon season. His feat was hailed the greatest achievement ever in mountaineering history. Messner chronicled the ascent in *The Crystal Horizon*. Without oxygen, he described the ordeal: "When I rest I feel utterly lifeless … I can scarcely go on. No despair, no happiness, no anxiety. I have not lost the mastery of my feelings, there are actually no more feelings. I consist only of will. After each few meters this too fizzles out in unending tiredness. Then I think nothing. I let myself fall, just lie there. For an indefinite time I remain completely irresolute. Then I make a few steps again." In 1986, concluding a difficult and demanding 16-year project, Messner summited the fourteenth and his last of the Himalayan peaks over 8000 m, the first to do so, all without supplemental oxygen.[3] His extraordinary accomplishments caused others to try Everest without oxygen. Some succeeded, but 10% died. Some who came through ended up with permanent mental impairments incurred from hypoxia.

Adaptations to Living at High Altitude

Aside from the adventurers who spend weeks or months preparing for high mountain ascents and those who sojourn less intensively for brief pleasure and recreation, about 500 million people, or 6% of the world's population, live and work at moderate altitude, 1500 m or above. About 80–140 million of them reside above 2500 m in the highlands and mountainous regions of Asia, South America, Mexico, and the western United States. Four South American capital cities are located at altitude, La Paz (3650 m), Quito (2850 m), Bogotá (2640 m), and Lima (1550 m). Roughly 60 settlements worldwide are established at 3500 m and above. About 18 of these (7 of them in Tibet) are higher than 4500 m, the highest being La Rinconada, Peru, at 5052 m, a gold mining community with 9746 inhabitants in its 2017 census. In 1913, a sulfur mine at 5950 m at the Aucanquilcha volcano in Chile was worked from a semi-permanent mining camp at 5300 m, with a network of roads leading up to the mine.

As people living and working in such areas have found, the higher the altitude the more challenging physical conditions become, and the harder it is to adapt and become self-sustaining. Temperature declines 6.5 °C for every 1000 m of altitude,

---

[3] In the 1990s, I heard Messner quip during a lecture, "If the meter were a little shorter, I would have had to climb more mountains."

thus if the temperature at sea level is 20 °C, expected values at 2000, 4000, and 6000 m would be progressively colder at 7, −6, and −19 °C. (Conditions are better at low latitudes because mean sea-level temperatures are warmer, and the atmosphere in the equatorial regions is thicker from the centrifugal force of Earth's rotation.) As altitude increases, suitable terrain for construction and habitation become scarce, topsoil is thin, little land is arable, and running water is limited. All these adversely affect people, livestock, agriculture, and native plant and animal life.

Altitude has a highly significant effect on human physiology. At sea level, the partial pressure of oxygen in the atmosphere, the corresponding arterial partial pressure of oxygen, and arterial hemoglobin oxygen saturation, are 159 mm Hg, 95 mm Hg, and 98%. But at 5500 m, the partial pressure in the atmosphere is only about half as much, 80 mm Hg. Correspondingly, in arteries it is only 50 mm Hg, and only 85% of hemoglobin is saturated (over the edge of "the slippery slope" of the hemoglobin-oxygen dissociation curve). At the summit of Mount Everest, the partial pressure of oxygen is only one-third that at sea level, 53 mm Hg, in the arteries it is only 32 mm Hg, and hemoglobin oxygen saturation is critical at 60%. (Recall for comparison that even mixed venous blood at sea level is 70% saturated.)

Upon rapid ascent to 1500–2500 m (the equivalent of flying, say, from sea level to Aspen, Colorado, altitude 2438 m), and particularly to over 2500 m, common symptoms even if only brief are fatigue, wooziness, shortness of breath, headache, loss of appetite, and tingling. We adapt by breathing more deeply or rapidly, which lowers the level of CO_2 in the alveoli and blood, providing a combined advantage of less oxygen displacement in the alveoli and a leftward curve shift (Bohr effect) resulting in better hemoglobin oxygen loading—at least until it is counterbalanced by increasing red cell 2,3-BPG production (a maladaptive response as it turns out). As additional adaptations, heart rate and cardiac stroke volume increase, increasing blood circulation to improve delivery of oxygen to tissues. And the low ambient oxygen environment activates the nitric oxide cycle that dilates blood vessels in the tissues for improved oxygen delivery. Almost immediately upon ascent, red cell production increases; each volume of blood can now carry more oxygen, and blood volume increases. After a brief time, most healthy individuals can adapt to altitudes up to 3500 m.[4]

However, altitude of this degree poses both acute and chronic risks for some people. Low ambient oxygen may cause blood vessels in the lungs to constrict. That, coupled with increased permeability of the smallest blood vessels, allows fluids to leak into the alveoli, causing cough and shortness of breath. This is "high-altitude pulmonary edema", commonly called HAPE for short. Another hazard is "high-altitude cerebral edema" (HACE); symptoms include lethargy, altered cognition, and nausea. Both are life-threatening disorders that require urgent

---

[4]Altitude greater than about 2500 m is inadvisable for people with important cardiopulmonary conditions, sickle cell disease, or a high-risk pregnancy.

supplemental oxygen and removal to lower altitude. Certain drugs are often used and helpful.[5]

A small number of individuals living or working at altitudes over 2500 m (usually higher) develop the syndrome of "chronic mountain sickness" (CMS). Reports vary on how frequently the syndrome occurs, but it seems to be just a few percent. The primary cause of the condition is extreme overproduction of red cells, which causes the blood to thicken so much (hyperviscosity) that it slows blood flow and causes pulmonary hypertension (back-up pressure in the lungs). The percentage of the blood that is red cells normally averages 42% in women and 45% in men, but in CMS values are typically over 57% and 63%, equating to hemoglobin levels greater than 19 and 21 g/dL (normal mean values being 14 and 15 g/dL). Blood volume is also markedly increased with reddish-blue skin and dilated veins. CMS has symptoms overlapping those of acute ascent, except that they persist. Sleep disturbances, sleep apnea, exhaustion, and mental problems are common. CMS is serious and can be life-threatening. Removal to significantly lower altitude is the only long-term solution, but short-term supplemental oxygen and removal of blood to reduce blood viscosity from all the red blood cells are helpful; drug therapies are not as predictably useful.[6] The occurrence of CMS will likely affect whether a person's family can continue to reside at high altitude. How and whether families who have lived at altitude for long periods may be pre-adapted by natural selection is unclear, but it is plausible that they are.

Older individuals and men are the more likely to develop the condition because arterial oxygen tension declines with age. C. A. Sorbini showed that mean low-altitude arterial partial pressure of oxygen of 95 mm Hg at age 20 falls to 75 mm Hg by age 75. F. León-Velarde and colleagues studied large numbers of long-term residents at three townships in Peru, Cojata (4355 m), Ananea (4660 m), and La Rinconada (given as 5500 m in their article, but actually 5052 m). A consequential proportion of men and smaller portion of women had hemoglobin levels greater than 21 g/dL, even as high as 23 g/dL, established determinants of CMS risk.

We now understand how the body increases red cell production in response to altitude. A connection between the kidneys and red cell production in the bone marrow was appreciated as long ago as the 1950s; and by the 1970s, erythropoietin (EPO) was identified as a mediator.[7] In the following decade, an explosion of research illuminated many key steps. A family of hypoxia-inducible factors (HIFs) is triggered whenever oxygen is in short supply. HIFs in turn activate the gene encoding EPO. Once the body's oxygen requirements are met, HIFs have to be

---

[5] For HAPE: the calcium channel blocker nifedipine, and phosphodiesterase-5 inhibitors tadalafil or sildenafil. For HACE: corticosteroids (dexamethasone) and carbonic anhydrase inhibitors (acetazolamide), both of which may be preventive as well.

[6] The carbonic anhydrase inhibitor (acetazolamide) is the best studied.

[7] That discovery led to the mass production of erythropoietin by recombinant technology to treat the anemia of renal failure starting in the 1980s.

degraded. That responsibility falls to a team of prolyl hydroxylase domain proteins and the von Hippel Lindau protein, along with additional modulators. Mutations encoding these genes may result in inappropriate red cell overproduction.[8]

Just as mutations in this pathway may lead to excess red cell production, it should come as no surprise that a mutation might do the opposite—impair the pathway and lead to underproduction of red cells despite ambient hypoxia. And such is the case. Whereas highlanders in the Andes rely on elevated red cell production to adapt to an oxygen-poor environment, circumstances are the opposite on the Tibetan plateau, a vast region of about 2.5 million square kilometers averaging an altitude of 4500 m, and home to about 5 million people. Separate studies by E. Huerta-Sánchez and J. Yang and their respective colleagues were revelatory. Most Tibetan highlanders have red cell counts and hemoglobin levels little different from those at low elevations, yet the highlanders function normally. How do they not develop marked increases in red cell count? They have a mutant gene for hypoxia-inducible factor 2α that does not properly sense the low-oxygen environment, thus impairing what should have been the typical EPO response. The frequency of this mutation in Tibetan highlanders is 87%, whereas it is 9% in the lowland Han Chinese from whom they split roughly 4700 years ago: the mutation was clearly highly adaptive and preserved at altitude. Intriguingly, it was present in the Denisovans, an ancient hominin group distinct from the Neanderthals that dwelt somewhat to the north of the Tibetan plateau across Asia. A Denisovan finger bone found in the Altai Mountains at 1600 m and a Denisovan molar tooth 160,000 years old found in the Tibetan highlands had the same mutation, strong indication of a common ancestral origin of these peoples and an ancient adaptation.

Still, how the Tibetan highlanders do so well without increased red cell production requires an explanation. A number of additional mutations distinct for the Tibetan highlanders were found in a genome-wide study: possibly one or more of them coevolved or cooperated to explain why they do not rely on higher numbers of red cells to improve their oxygen delivery. Tibetans do have marked elevations in hemoglobin S-nitrosothiol to enhance tissue blood flow, and this may be one explanation. Mutations governing prolyl hydroxylase domain protein expression and muscle mitochondrial function appear to be others. (Present-day Himalayan Sherpas, renowned for their tolerance to altitude, are descendants of Tibetan highlanders going back about 500 years, recently enough that the genetic makeups of the two populations should be similar.)

---

[8] One of them distinguished itself because it is so serious, "Chuvash polycythemia", found in the Chuvash Republic of Russia, due to a VHL mutation inherited from both parents. As a result, unopposed HIFs result not only in red cell overproduction but also in pulmonary hypertension and thrombosis.

The Equivalent of High Altitude at Low Altitude

A low-altitude equivalent of living at very high altitude is cyanotic congenital heart disease (CCHD). Affected individuals are born with abnormal heart anatomy such that some of the venous blood returning to the heart and lungs for reoxygenation bypasses the lungs. Thus blood being pumped out through the major arteries is a mixture of oxygenated and deoxygenated blood. CCHD resembles living at high altitude. The arterial hemoglobin-oxygen saturation in CCHD ranges from about 60% to 90%. In a study published in 1986 that I conducted, mean red cell mass in CCHD patients was about twice normal and total blood volume half above normal. Whereas normal individuals are 7% blood, CCHD patients are about 10.5% blood. The P50 shifted slightly rightward. Most CCHD individuals found an optimal red cell equilibrium where stable, increased red cell production satisfied needs. But some did not: in them, red cell production proceeded unabated and hyperviscosity ensued, a maladaptive extreme called "decompensated erythrocytosis", with para-doxically worsening shortness of breath and mental functioning. Red cells could be made so rapidly that there was not enough time for the developing red cell to get its full complement of iron for hemoglobin production. These CCHD patients are simi-lar to high-altitude residents with chronic mountain sickness. With CCHD, how-ever, oxygen supplementation is largely ineffective because it cannot reach blood bypassing the lungs although it might maximize saturation in blood that does flow through the lungs, and a very small amount can dissolve in blood plasma. Drugs that relieve pulmonary hypertension may in theory be able to redirect some of the blood flow toward the lungs. The longstanding principal treatment has been removal of blood to relieve hyperviscosity. That inevitably leads to iron depletion, a counter-productive result. Most patients undergo surgery at a young age to attempt to fix the abnormalities, but surgery may be only partly or impermanently corrective.

Similar to CCHD, there are rare patients who have vascular connections bypass-ing the lungs. And in another scenario, some patients with chronic lung diseases, despite their use of supplemental oxygen, still will not achieve hemoglobin-oxygen saturations above 85–90%. In them, EPO drives increased red cell production.

At Altitude, Which Way Should the Hemoglobin-Oxygen Dissociation Curve Shift?

In the anemias, and in sea-level dwellers at altitude, the hemoglobin-oxygen disso-ciation curve shifts to the right, due to stepped-up 2,3-BPG production. Therefore, "common wisdom" has had it that a rightward shift at altitude is an important adap-tation because it would facilitate unloading of oxygen at the tissue level. However, a number of observations have called this assumption into question, headed by one fundamental reality—at ever-increasing altitude a rightward shift progressively compromises oxygen loading in even healthy lungs.

Mairbäurl and Weber adjusted human hemoglobin in the laboratory to P50s of 24, 26.8, or 30 mm Hg, that is, modestly left-shifted, normal, and right-shifted. With sea level oxygen tension, the three hemoglobins were all well oxygen saturated, 95% or greater. However, with an oxygen tension expected at 4500 m, the three respective saturations were now substantially different between them; the leftward shift was best for oxygen loading at 85.2% versus 80.7% and 75.2%, respectively. These would correspond to respective altitudes of 3500 m, 4500 m, and 5500 m. The left shift would represent a very substantial advantage at altitude.

The same left shift adaptation has been documented in other animals. A number of bird species achieve extraordinary altitudes in flight. A Rüppell's Vulture (*Gyps rueppelli*) was recorded at 37,000 feet (11,278 m) by an airline pilot who glanced at his altimeter when he collided with the bird. The Bar-headed Goose (*Anser indicus*) flies over the Himalayas during migration at altitudes over 8000 m. The Andean Goose (*Chloephaga melanoptera*) spends virtually its entire life in the Andes Mountains at elevations above 3000 m. It so happens they all have left-shifted hemoglobin-oxygen dissociation curves. The bottom-dwelling bony fish *Pleuronectes platessa* naturally lives in hypoxic conditions and has a markedly left-shifted curve that shifts rightward only when the fish's environment is more favorably oxygenated.

The four camelids of South America, the vicuña (*Vicugna vicugna*), llama (*Lama glama*), alpaca (*Lama pacos*), and guanaco (*Lama guanicoe*) thrive at altitude. All have hemoglobins differing among them by various amino acid substitutions that downregulate 2,3-BPG binding. Their P50s are thereby left shifted compared to most other mammals. The most left shifted is the vicuña with a P50 of 17.5 mm Hg; vicuñas live at the highest altitude among the camelids, about 3200–4800 m. The llama and alpaca (now domesticated species) both have P50s of 20.3 mmHg, while the guanaco that lives at lower altitude, has a P50 of 22.2 mm Hg. These left shifts confer high oxygen loading in the lungs despite the hypoxic atmosphere. Vicuñas and llamas do not increase red cell production because they presumably do not need to. The guanaco, however, with the least left-shifted P50, does increase its red cell production when living at the high end of its range, about 4000 m. The mutations that resulted in the camelids' left-shifted P50s, like all mutations, would have been spontaneous and unintentional. However, these shifts would have pre-adapted them. All four are herbivorous and are grazers, and perhaps the vicuña's especially marked left shift helped it occupy the most altitudinous niches for grazing, where there would be reduced competition for scarce food sources.

In a fascinating and elucidating 1978 paper titled "Human llamas", Herbel and colleagues identified a family with four teenagers, two (one male, one female) with a high-affinity hemoglobin variant, named Hb Andrew-Minneapolis, in which the amino acid asparagine substituted for lysine at position 144 in the β-globin chain. The P50 values of the teens with the variant were markedly left shifted at 17.0 and 17.2 mm Hg. The two normal siblings (one male, one female) had normal P50s of 26.9 and 27.0 mm Hg. In Minneapolis (altitude 245 m), the mutated siblings had elevated hemoglobin levels of 16.8 and 17.3 g/dL, whereas the normal ones had normal values of 14.1 and 13.5 g/dL. The left-shifted hemoglobin of the affected siblings was not able to deliver oxygen adequately to their tissues, so they ramped up red cell production to compensate. But even so, when their oxygen consumption was tested on a stationary bicycle ergometer, the mean maximal amount they could take in and use was lower than that of their normal siblings, 34.9 vs. 43.0 mL/kg/min.

The story at high altitude was the reverse. The four siblings were taken to Leadville, Colorado, at altitude 3100 m. This time, the bicycle ergometer showed that the two affected siblings' mean maximal oxygen consumption *increased* despite the altitude gain, from 34.9 to 39.5 mL/kg/min, whereas the two normal siblings' maximal oxygen consumption dropped markedly, from 43.0 down to 32.6 mL/kg/min. In other words, the siblings with the mutations were pre-adapted to a low-oxygen environment. They were just like llamas, or even more like vicuñas, given their P50 values. If they chose to live at high altitude, they would have a genetic advantage in oxygen delivery to tissues. However, a pregnant woman with the mutation would have a left-shifted P50 even lower than fetal hemoglobin, 19 mm Hg; oxygen transfer to the fetus would be compromised. (But a surrogate could carry the pregnancy after egg procurement and in vitro fertilization.)

All these compelling lines of evidence support the advantage of a left shift under hypoxic conditions, thus begging the question why ascent to altitude results in a rightward shift instead. One hypothesis is that in ancient times, dwelling at altitude was probably infrequent for humans, but anemias of diverse causes, especially iron deficiency and blood loss, were surely very common. A rightward shift as a response to anemia at low altitude would be highly adaptive because oxygen loading in the lungs would remain excellent while the shift would promote easier oxygen release to the tissues. Possibly our species therefore defaulted to a rightward shift to any kind of oxygen deprivation. But at altitude we would depend on increased numbers of red cells (with their hemoglobin) and hope not to develop pulmonary hypertension and chronic mountain sickness. If you were the equivalent of a vicuña born with a rare left-shifted hemoglobin variant, or you had a condition of very high fetal hemoglobin, or you were a Tibetan highlander, you would be pre-adapted.

Observations and Lessons from Unusual Mutant Hemoglobins

In all, well over a thousand variant hemoglobins in humans have been described, all of them rare except those responding to malaria. Most are functionally neutral, but over one hundred demonstrate altered oxygen affinity. The aforementioned "human llama" Hb Andrew-Minneapolis is one of them. The structural and functional details of many of these have been elegantly characterized. (Inheritance from just one parent suffices to produce a functional change.) At low elevation, variants that shift the hemoglobin-oxygen dissociation curve leftward may stimulate increased red cell production. Those that shift it rightward, by making oxygen more readily available, may do the opposite, fewer red cells sufficing and producing the false impression of anemia in routine lab tests.

For readers who would like to take a closer look, two examples of abnormal oxygen affinity illustrate these types. In 1975, Jensen and colleagues described an extremely left-shifted hemoglobin-oxygen dissociation curve affecting six people in a family over four generations. They functioned well but had markedly increased hemoglobin levels ranging from 19.2 to 23.8 g/dL. The cause was a single amino acid substitution of proline for histidine at position 143 in the β chain, named Hb Syracuse. Proline resists twisting into the helix formation in hemoglobin's secondary structure; one segment of the helix was shortened one amino acid, changing the hemoglobin's tertiary and quaternary structures. The change at position 143 also impaired an important 2,3-BPG binding site. Hb Syracuse thus manifested a triple fault of poor 2,3-BPG binding, poor Bohr effect, and disordered cooperativity. The P50 was astoundingly left-shifted to 11 mm Hg (normal is 27 mm Hg). (Of note is that fetal hemoglobin also has a substitution at position 143; its serine similarly impairs 2,3-BPG binding, contributing to the evolutionarily intended left-shifted P50. But the serine substitution does not shorten that segment of the helix, and therefore does not interfere with the Bohr effect or cooperativity.) (Fig. 9.1).

An example of an extremely right-shifted hemoglobin was described by Reissmann in 1961 and further elucidated by Bonaventura in 1968. Its cause is a single amino acid substitution of threonine for asparagine at position 102 of the β-globin chain, named Hb Kansas. The amino acid switch causes abnormalities in the molecule's tertiary and quaternary structures. The original case, a 14 year-old teenager and his mother both had cyanosis, a bluish cast to the skin from a high level of deoxygenated hemoglobin; but they were otherwise well except fatigue on exertion. The teen had a normal hemoglobin level of 13.9 g/dL. The P50 was extremely right shifted at 70 mm Hg. His lungs were normal, and he had an excellent, even supranormal, arterial oxygen tension of 96–104 mmHg at which adult hemoglobin

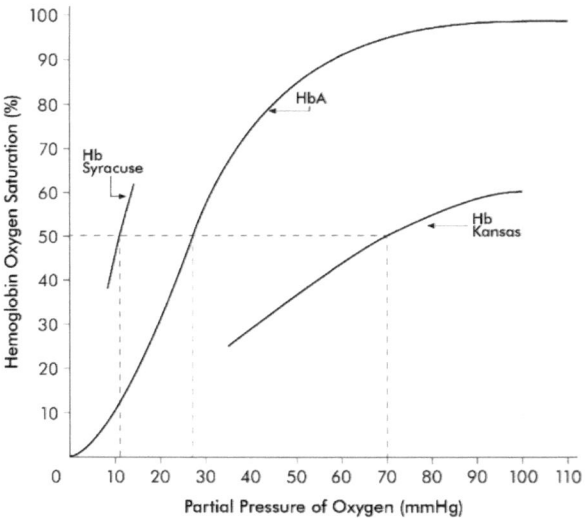

Fig. 9.1 The hemoglobin-oxygen dissociation curves for HbA, Hb Syracuse, and Hb Kansas

would normally be 98% or more oxygen saturated—but his was only 60%, because his hemoglobin had such low affinity for oxygen. His 40% deoxygenated hemoglobin caused the physical appearance of cyanosis, present when, from any cause, it exceeds 4 g/dL, which his did. His symptoms were minimal, however, because at a mean tissue oxygen tension of 35 mm Hg, which is about normal, his oxygenated hemoglobin was 25%, meaning he successfully unloaded 35% of the oxygen being carried, similar to what adult hemoglobin does. Individuals with right-shifted mutations might benefit athletically because oxygen unloading in exercising muscles would be facilitated—but only if oxygen loading in the lungs were not seriously compromised, which his was.

Some hemoglobins cannot maintain heme iron in the ferrous (Fe^{2+}) state. These are the congenital methemoglobinemias, in which a portion of the iron is in the ferric state (Fe^{3+}), incapable of binding oxygen. Affected individuals may have elevated hemoglobin levels to compensate. The degree of methemoglobinemia is variable and often without symptoms or signs, except the bluish skin of cyanosis when the level exceeds 1.5 g/dL. (Methemoglobin produces cyanosis more easily than deoxygenated hemoglobin.)

The commoner among two principal causes is deficiency of the red cell enzyme cytochrome b5 reductase (Cyb5R), which constantly works restoring Fe^{3+} back to Fe^{2+} during the oxidative stresses of daily life. Since one normal gene is sufficient for normalcy, individuals with methemoglobinemia have two abnormal genes; consanguinity (meaning the parents are related, first cousins for example) is common. Affected individuals worsen if exposed to oxidizing drugs. When blood is taken for testing, it may appear somewhat chocolate brown. The arterial oxygen tension is normal because lung function is normal, but arterial oxygen saturation is low (because affected hemes cannot bind oxygen). This incongruity is a valuable clue to diagnosis. For nearly all affected individuals, Cyb5R deficiency is cosmetic only. But for those concerned about their appearance or who may have symptoms, drugs that can reduce Fe^{3+} to Fe^{2+} can be helpful, including methylene blue or high-dose vitamin C.

Rarer causes of methemoglobinemia are hemoglobin mutations that destabilize ferrous iron (Fe^{2+}), collectively grouped as "hemoglobins M" (HbM). Affected individuals need only inherit the abnormal gene from one parent. Histidine residues anchor heme, so a mutation that substitutes a tyrosine for histidine near heme explains nearly all cases. The time frame of appearance of cyanosis in life depends on whether the α, β, or γ subunit is affected. If α, cyanosis is evident from birth and persists. If γ, the condition disappears as fetal gives way to adult hemoglobin. And if β, the abnormality appears during the first half year of life when adult hemoglobin takes over from fetal hemoglobin. Methylene blue and vitamin C confer no benefit. The condition for most is only a cosmetic nuisance.

Blood Doping

We have looked in this chapter at various kinds of circumstances that limit oxygen availability and how we adapt to them—high altitude, cardiopulmonary disorders, anemia and genetic abnormalities of hemoglobin. I will close this chapter by saying a few words about some athletes' deliberate attempts to increase oxygen availability for the sole purpose of enhancing aerobic, athletic performances through boosting red cell numbers or changing hemoglobin function. Collectively these activities are referred to as "blood doping". It is a fact that oxygen availability is performance-limiting in the most aerobic sports including middle- and long-distance running, cycling, swimming, and cross-country skiing. An athlete's body is not inherently oxygen deprived, but athletes who desire as much aerobic performance as possible may perceive or come to believe they have a relative "deficiency" that they have the need or right to correct.

Doping succeeds in raising oxygen transport and delivery, but it is dangerous owing to the risk of cardiovascular complications and blood clots. On the broader social level, it puts athletes who do not engage in such activities at a competitive disadvantage, and it results in artificial, new national, world, and Olympic records. It is a serious breach of ethics. Fortunately, doping, which was common if not widespread until the last decade or so, has since been largely controlled by internationally recognized policies under the World Anti-Doping Code that has defined which substances and practices are banned and what mandatory testing is done.[9]

Further Reading

[Anon.] List of highest settlements. https://en.wikipedia.org/wiki/List_of_highest_settlements.

Beall CM, Cavalleri GL, Deng L, Elston RC, Gao Y, Knight J, et al. Natural selection on EPAS1 (HIF2alpha) associated with low hemoglobin concentration in Tibetan highlanders. Proc Natl Acad Sci USA. 2010;107:11459–64. https://doi.org/10.1073/pnas.1002443107.

Benz EJ. Hemoglobin variants associated with hemolytic anemia, altered oxygen affinity, and met-hemoglobinemias. In: Hoffman R, et al., editors. Hematology: basic principles and practice. 3rd ed. New York: Churchill Livingstone; 2000. p. 554–61.

Beutler E. Hemoglobinopathies associated with unstable hemoglobin. In: Beutler E, et al., editors. Williams Hematology. 6th ed. New York: McGraw-Hill; 2001. p. 607–10.

---

[9] The U.S. Anti-Doping Agency, a signatory to the Code, is a non-governmental agency recognized by the U.S. Congress. Banned drugs include the following: drugs already in use or under study in people with anemia due to chronic kidney diseases including the erythropoiesis-stimulating agent erythropoietin and the longer-acting darbepoietin; drugs that stabilize hypoxia-inducible factors by inhibiting prolyl hydroxylases, such as daprodustat, roxadustat, and vadadustat; myo-inositol tri-spyrophosphate that shift the hemoglobin-oxygen dissociation curve rightward which improves oxygen unloading to muscles at low altitudes; numerous anabolic steroids that raise hemoglobin levels and increase muscle bulk; transfusion of red cells for the sake of athletic performance, whether one's own red cells or another's; and numerous other performance-enhancing drugs unrelated to hemoglobin.

Bert P. La pression Barométrique: Recherches de Physiologie Expérimentale. Paris: G. Masson; 1878.

Bonaventura J, Riggs A. Hemoglobin Kansas, a human hemoglobin with a neutral amino acid substitution and an abnormal oxygen equilibrium. J Biol Chem. 1968;243:980–91. https://doi.org/10.1016/S0021-9258(18)93612-4.

Dominguez de Villota ED, Ruiz Carmona MT, Rubio JJ, de Andrés S. Equality of the in vivo and in vitro oxygen-binding capacity of hemoglobin in patients with severe respiratory disease. Br J Anaesth. 1981;53(12):1325–8. https://doi.org/10.1093/bja/53.12.1325.

Eaton JW, Skelton TD, Berger E. Survival at extreme altitude: protective effect of increased hemoglobin-oxygen affinity. Science. 1974;183:743–4. https://doi.org/10.1126/science.183.4126.743.

Erslev AJ. Blood and mountains. In: Wintrobe MM, editor. Blood pure and eloquent: a story of discovery, of people, and of ideas. New York: McGraw-Hill; 1980. p. 257–80.

Hackett PH, Roach RC. High-altitude illness. N Engl J Med. 2001;345:107–14. https://doi.org/10.1056/NEJM200107123450206.

Hall FG, Dill DB, Barron ESG. Comparative physiology in high altitudes. J Cell Comp Physiol. 1936;8:301–13. https://doi.org/10.1002/jcp.1030080302.

Herbel RP, Eaton JW, Kronenberg RS, Zanjani ED, Moore LG, Berger EM. Human llamas. J Clin Invest. 1978;62:593–600. https://doi.org/10.1172/JCI109165.

Huerta-Sánchez E, Jin X, Asan, et al. Altitude adaptation in Tibet caused by introgression of Denisovan-like DNA. Nature. 2014;512(7513):194–7. https://doi.org/10.1038/nature13408.

Jensen M, Oski FA, Nathan DG, Bunn HF. Hemoglobin Syracuse ($\alpha_2\beta_2^{143(h21)His \rightarrow Pro}$), a new high-affinity variant detected by special electrophoretic methods. J Clin Invest. 1975;55:469–77. https://doi.org/10.1172/JCI107953.

León-Velarde F, Gamboa A, Chuquiza JA, et al. Hematological parameters in high altitude residents living at 4,355, 4,660, and 5,500 meters above sea level. High Alt Med Biol. 2000;1:97–104. https://doi.org/10.1089/15270290050074233.

León-Velarde F, Maggiorini M, Reeves JT, et al. Consensus statement on chronic and subacute high-altitude diseases. High Alt Med Biol. 2005;6:147–57. https://doi.org/10.1089/ham.2005.6.147.

Luks AM, Hackett PH. Medical conditions and high-altitude travel. N Engl J Med. 2022;386:364–73. https://doi.org/10.1056/NEJMra2104829.

Krakauer J. Into thin air: a personal account of the Mount Everest disaster. New York: Villard; 1997.

Mairbäurl H, Weber RE. Oxygen transport by hemoglobin. Compr Physiol. 2012;2:1463–89. https://doi.org/10.1002/cphy.c080113.

Messner R. The crystal horizon: Everest—the first solo ascent. Seattle: Mountaineers Books; 1989.

Perutz M. Molecular anatomy, physiology, and pathology of hemoglobin. In: Stamatoyannopoulos G, et al., editors. Molecular basis of blood diseases. Philadelphia: W. B. Saunders; 1987. p. 127–78.

Reissmann KR, Ruth WE, Nomura T. A human hemoglobin with lowered oxygen affinity and impaired heme-heme interactions. J Clin Invest. 1961;40:1826–33. https://doi.org/10.1172/JCI104406.

Rosove MH, Perloff JK, Hocking WG, Child JA, Canobbio MM, Skorton DJ. Chronic hypoxaemia and decompensated erythrocytosis in cyanotic congenital heart disease. Lancet. 1986;2(8502):313–5. https://doi.org/10.1016/s0140-6736(86)90005-x.

Roxie C. Collision between a vulture and an aircraft at an altitude of 37,000 feet. Wilson Bull. 1974;86(4):461–2.

Semenza GL. The genomics and genetics of oxygen homeostasis. Annu Rev Genomics Hum Genet. 2020;21:183–204. https://doi.org/10.1146/annurev-genom-111119-073356.

Simonson TS, Yang Y, Huff CD, et al. Genetic evidence for high-altitude adaptation in Tibet. Science. 2010;329:72–5. https://doi.org/10.1126/science.1189406.

Sorbini CA, Grassi V, Solinas E, Muiesan G. Arterial oxygen tension in relation to age in healthy subjects. Respiration. 1968;25:3–13.

Storz JF. Hemoglobin function and physiological adaptation to hypoxia in high-altitude mammals. J Mammal. 2007;88:24–31. https://doi.org/10.1644/06-MAMM-S-199R1.1.

Tremblay JC, Ainslie PN. Researchers re-evaluate estimate of the world's high-altitude population. Proc Natl Acad Sci USA. 2021;118:e21024631. https://doi.org/10.1073/pnas.2102463118.

Wood SC, Johansen K, Weber RE. Effects of ambient P_{O2} on hemoglobin-oxygen affinity and red cell ATP concentrations in a benthic fish, Pleuronectes platessa. Respir Physiol. 1975;25:259–67. https://doi.org/10.1016/0034-5687(75)90002-x.

West J, Boyer SJ, Graber DJ, et al. Maximal exercise at extreme altitudes on Mount Everest. J Appl Physiol. 1983;55:688–98. https://doi.org/10.1152/jappl.1983.55.3.688.

West JB. Highest permanent human habitation. High Alt Med Biol. 2002;3:401–7. https://doi.org/10.1089/15270290260512882.

Yang J, Jin Z-B, Chen J, et al. Genetic signatures of high-altitude adaptation in Tibetans. Proc Natl Acad Sci USA. 2017;114:4189–94. https://doi.org/10.1073/pnas.1617042114/-/DCSupplemental.

Chapter 10
The Porphyrias

Up to this point, I have presented heme only as an integral part of hemoglobin doing its job of binding and releasing oxygen and participating in the nitric oxide cycle. However, it serves additional functions in humans as an essential component of the cytochrome system in liver cells responsible for day-to-day metabolic functions and as a component of myoglobin and cofactor for a number of enzymes.

In this chapter, I'll focus on a gallery of maladies that result when the synthesis of heme goes awry. All these conditions are purposeless, all of them are nasty, and fortunately, all of them are quite rare. They are called the porphyrias, named for heme's basic porphyrin chemical structure.

Patients with the Classic Porphyria Syndromes

The following four patients are typical of these disorders:

Patient #1: A woman in her late teens had never before had a medical problem. Then, in a 6-month stretch, she had abdominal pains, cramping, and occasional vomiting. Her symptoms sometimes eased up, only to recur. They were worse after eating and especially just before menstrual periods. She lost 30 pounds. She was seen by numerous doctors, hospitalized several times, and had myriad blood and urine tests (but not the right ones) plus scans and endoscopies that showed nothing. As her pain and anxiety increased, she needed strong medications, and one physician suggested she had a drug dependence problem and recommended a psychiatrist. Then, over a few days, matters became critical. She lost most of the strength in her arms and legs, had difficulty swallowing and urinating, and had a brief seizure. This time she went to a different hospital, crying in pain. A fresh team of doctors proposed a rare diagnosis for the first time, urine tests put on rush were positive, and treatment for her particular porphyria was started right away. She was quickly on the road to improvement.

Patient #2: A woman in her middle years had been sensitive to sunlight since her earliest childhood. She would develop painful sunburns and figured out on her own to avoid the sun and wear protective clothing, since UVA/UVB sunblocks were ineffective. Over-the-counter pain medications were of no help in easing the pain of her burned skin. She could only be outside with her children on very cloudy days, and with so little sun exposure, her vitamin D level was low and she required a supplement. Her new primary care doctor referred her to a dermatologist who thought of testing her for one of the porphyrias. Specialized tests yielded the diagnosis, and routine blood tests showed minor liver abnormalities. She was started on treatment and could now enjoy her children in almost all outdoor circumstances, although her liver problem persisted and remained an ongoing concern.

Patient #3: A man in his late middle years had had almost no medical care because of an innate aversion to doctors. For a number of years he sunburned easily, with blisters on the face and the backs of his forearms and hands, which left scarring and pigmentation. His skin condition finally aroused his concern, even more so his wife's, who wondered what it meant and did not like its appearance. He was a modest drinker, and his blood tests showed he was loaded with iron and had hepatitis C virus. Urine testing was diagnostic, and treatments were started with good results.

Patient #4: A 2-year-old girl whose parents were first cousins had been anemic from birth and required frequent red cell transfusions. Her skin was extremely sensitive to sunlight, and even her eyelids blistered with mild sun exposure. Her teeth and urine had a reddish-brown discoloration. Her porphyria was diagnosed with blood and urine tests shortly after birth, and her parents were strongly advised to keep her indoors as much as possible and provide a vitamin D supplement. But finally, because of the child's unremitting skin sensitivity and anemia, the parents were advised to seek allogeneic bone marrow stem cell transplantation for their daughter at a university hospital. After grappling with the emotional turmoil knowing they were subjecting her to a risky procedure, they agreed to proceed. All went well, and the anemia and skinburning resolved.

The General Character and Challenges of the Porphyrias

When heme is assembled—step by step in eight stages—a different enzyme is involved in completing each. If one of the enzymes cannot complete its step, heme synthesis is derailed: the amount of heme produced is reduced, and heme precursors behind the blocked enzyme accumulate. Each kind of porphyria is the result of one deficient enzyme specific to either liver cells or red cells. When the precursors are released, they are noxious, producing a range of symptoms and problems. All but one are predominantly genetic and run in families, but because manifestations in family members are so variable, a connection may not be obvious.

Different precursors attack different parts of the body—variably the nervous system, skin (in combination with sunlight), red cells, and liver cells. The clinical problem depends on which enzyme is deficient and whether the heme precursors build up primarily in liver or red cells. While most patients fit fairly neatly into one of the few aforementioned syndromes, in some cases syndromes overlap.

Diagnosis can be difficult, but not because specific blood and urine tests aren't readily available in reference laboratories. The problem is the collective rarity of the porphyrias. Most physicians are only vaguely familiar with them, and they are virtually unknown to the general public. Because patients' complaints may have various explanations, a physician might not think of porphyria even though the diagnosis would be easy if appropriate tests were done. On the other hand, physicians sometimes reach for porphyria when faced with inexplicable symptoms; and test results falling minimally outside the normal ranges may lead to over-diagnosis. Most patients referred to hematologists for porphyria in fact do not have it, but those who do pose important and often difficult treatment challenges.

Complicating matters, there is no uniformity in the way physicians refer to the different conditions. It would make sense to name the porphyria by its defective enzyme—but some enzymes have more than one legitimate name, and many of them are a hard-to-remember jumble of syllables. Naming a porphyria for the specific DNA mutation that causes it would be impractical and unhelpful because the enzyme deficiency is what matters clinically, and each enzyme deficiency may be explained by any number of mutations. So it has been customary to revert to the popular descriptive names that evolved long before the genetics and chemistry were understood—only one porphyria is actually named for the responsible enzyme deficiency.

Next, we'll take a closer look at these four patients and what goes on at the level of the specific enzyme problems and precursor accumulations in the heme synthesis pathway, and at which treatments can be helpful. The subject matter is admittedly difficult. Much of the following may be oriented toward physicians, health care providers, and affected patients and their families, but if you are game, read on!

The "Acute Hepatic Porphyrias"—The Neurovisceral Syndrome

Patient #1, a young lady with prominent neurological and gastrointestinal symptoms, had what is called the "neurovisceral" syndrome. All manifestations result from toxic effects on nerves. Abdominal symptoms, which are nearly always present, are caused by dysfunction of the autonomic nervous system that may also cause a faster-than-normal heart rate and high blood pressure. Neural dysfunction can affect the brain, nerves to the arms and legs, and internal structures. In the extreme, it can paralyze breathing. Abnormal water retention (cued by a low serum sodium level) is common. The first two precursors on the pathway to heme synthesis, respectively delta-aminolevulinic acid (δ-ALA) and porphobilinogen (PBG), are responsible. They build up when they are logjammed on account of an enzyme deficiency downstream, but how they are neurotoxic is poorly understood.

The commonest explanation is deficiency of the third enzyme in the heme synthesis chain, PBG deaminase. The condition's common name is "acute intermittent porphyria". Less frequent causes are deficiency of either the sixth or seventh enzyme, coproporphyrinogen oxidase or protoporphyrinogen oxidase, respectively called "hereditary coproporphyria" and "variegate porphyria", which may also be accompanied by sensitivity to sunlight. And rarest of all, deficiency of the second enzyme, δ-aminolevulinic acid hydratase, is named after the enzyme itself; in that disorder only δ-ALA is in excess as PBG is downstream from the defective enzyme. Various blood and urine tests can readily sort these out for the patient and testing first-order family members. Deficiency in liver cells is responsible, not red cells, earning these conditions as a group an additional appellation, the "acute hepatic porphyrias". Liver cell cancer is a rare complication.

Just as important as the enzyme deficiency itself is that activity of the first enzyme, δ-ALA synthase 1 (δ-ALAS1), the enzyme form specific to the liver, ramps up because there is not enough heme to provide feedback and shut it down normally. That results in even more δ-ALA and PBG, a double whammy. Sometimes there is so much PBG that a breakdown product of it imparts a reddish-brown color to the urine. δ-ALAS1 is the logical and principal therapeutic target to cut down δ-ALA and PBG production.

Hemin, a modified heme made from human blood, is a highly effective treatment—it substitutes for natural heme to downregulate δ-ALAS1. For long-term management, the drug givosiran is more practical: it directly suppresses the enzyme and avoids what would be the inevitable accumulation of iron from numerous hemin treatments. However, some questions remain regarding its usage, particularly its effect on other heme-dependent liver enzymes that may, for example, affect drug dosing. Some patients can identify aggravating factors to avoid, such as low food or carbohydrate intake. Some women with pre-menstrual neurovisceral attacks are helped by drugs that suppress the menstrual cycle (but avoiding progesterone that stimulates δ-ALAS1). Various drugs have been found to aggravate the condition, and they must be avoided even though it may not be understood how they provoke it. In extreme cases, liver transplantation, for which liver cell cancer is often the indication, can cure both the cancer and porphyria at the same time. The American Porphyria Foundation (porphyriafoundation.org) and the International Porphyria Network (porphyrianet.org) provide valuable information on all these and the rest of the porphyrias.

The Red Blood Cell "Protoporphyrias"—Sunburn and Liver Problems

Patient #2, a middle-aged woman, had painful sunburn without blistering or skin damage, and also had concerning abnormal liver tests from an early age. That constellation of findings is typical of "erythropoietic protoporphyria", which is caused by deficiency of the last, eighth enzyme, ferrochelatase. To produce a clinical problem, a gene programming for a defective enzyme must be inherited from both parents. Red cell ferrochelatase, as the name suggests, causes protoporphyrin to chelate into its structure an iron atom to complete heme synthesis. (Ferrochelatase may alternatively insert zinc.) When ferrochelatase is deficient, metal-free (that is, zinc-free) protoporphyrin builds up in red cells to very high levels. The diagnosis is straightforward by measuring total and metal-free protoporphyrin.[1]

During the course of a red cell's lifespan, metal-free protoporphyrin gradually leaves the red cell and binds to circulating lipids and proteins. The complexes lodge in the skin, where sunlight or artificial light sources cause reactive oxygen species to form. That activates the immune system, damaging the cells lining the inner blood vessel wall. Affected people find that UVA/UVB sunblocks don't protect their skin. That is because these sunblocks absorb wavelengths in the ultraviolet range of about 280–400 nm, whereas protoporphyrin complexes are activated at slightly longer violet wavelengths, 400–420 nm. That bandwidth can be screened out by zinc-based sunscreens or by enhancing skin melanin deposition from treatment with the FDA-approved drug afamelanotide. Hemin is not helpful because it does not down-regulate the δ-ALAS2 enzyme specific to red cells.

Protoporphyrin-lipoprotein complexes in the blood also cause liver trouble. They are taken up by the liver, enter the biliary system, damage bile ducts, and commonly cause gallstones. When protoporphyrin in bile reaches the intestinal tract, it can be reabsorbed into the circulation causing damage all over again; drugs that interfere with that recycling can be helpful. A few percent of patients experience progressive liver failure and may require liver transplantation, but protoporphyric disease will recur in the transplant unless the source of excess protoporphyrin is eliminated by allogeneic bone marrow stem cell transplantation. Stem cell transplantation can be used preemptively when it is obvious that the liver is on a path to failure. Experimental drug therapies are being evaluated. Gene therapies should eventually be feasible and would be ideal.

An additional protoporphyric disorder is "X-linked protoporphyria". This is a unique porphyria that results not from an enzyme deficiency, but rather from a *gain-of-function* mutation in the first enzyme in red cells, δ-ALAS2. (Gain of function means the mutation programs for *increased* enzyme activity.) The gene is on the X chromosome, hence the name. As affected males (XY) have only one X, all enzyme they make is deficient and porphyria is significant, whereas women (XX) with one abnormal X and one normal X have a milder condition.

In X-linked protoporphyria all downstream heme precursors starting with δ-ALA are overproduced, but only the buildup of red cell protoporphyrin is usually consequential. Ferrochelatase (the eighth enzyme) is normal but overwhelmed by the upstream delivery of a large protoporphyrin load. The clinical picture is virtually the same as in Patient 2's erythropoietic protoporphyria due to ferrochelatase deficiency, but X-linked protoporphyria is milder. Clinical management is similar.

---

[1] Iron deficiency or lead poisoning both raise total protoporphyrin levels, but zinc protoporphyrin rather than metal-free protoporphyrin is elevated because ferrochelatase is normal.

Whereas in ferrochelatase deficiency protoporphyrin is predominantly metal-free for lack of enzyme activity, in X-linked protoporphyria ferrochelatase is normal and therefore can generate zinc protoporphyrin, a point of diagnosis that separates the two entities.

Porphyria Cutanea Tarda, the "Chronic Hepatic Porphyria"—Sunburn and Blisters

Patient #3, a man in his late middle years, had blistering, scarring skin lesions in sun-exposed areas of the skin over a prolonged period of time typical of "porphyria cutanea tarda", which literally means "porphyria affecting the skin later in life". It appears to be the commonest of these disorders. The fundamental cause is reduced activity of the fifth enzyme in the heme synthetic pathway, uroporphyrinogen decarboxylase, specifically in the liver. Given the long time frame of the condition, it has been referred to as "chronic hepatic porphyria". The heme precursor uroporphyrinogen III and its derivatives pile up just in front of the enzyme block. They filter into the urine and may discolor it wine red. They also bring about the photosensitivity and skin damage. Whereas other porphyrias are due predominantly to DNA mutations, uroporphyrinogen decarboxylase is notably susceptible to non-genetic factors, meaning this porphyria is as much or more dependent on these than genetic ones. The condition often develops when there is iron accumulation in the liver, as in unrelated congenital disorders of excessive iron absorption or otherwise unrelated chronic liver diseases such as hepatitis C. Depleting total body iron is highly therapeutic. That is the first-line therapy: patients are helped by removing blood repeatedly until they are at the verge of frank iron deficiency. Hepatitis C virus is usually treated with anti-viral medication. Alcohol plays a minor role but consumption is best reduced or eliminated if hepatitis C or liver disease is present.

"Congenital Erythropoietic Porphyria"—Photomutilation and Red Blood Cell Destruction

Patient #4, a young girl, had one of the rarest disorders, and arguably the most vicious, "congenital erythropoietic porphyria", caused by deficiency of the fourth enzyme, uroporphyrinogen synthase. Clinical problems occur when one inherits an abnormal gene from both parents. Consanguinity (meaning that the parents are members of the same family) is common, as in this patient's case. Red cells are the source of the problem, not liver cells. The heme synthesis enzyme blockade in this case leads to buildup of the heme precursor hydroxymethylbilane, which in turn is converted into non-usable, derivative toxic porphyrins that damage red cells. Porphyrins liberated from red cells damage the skin and discolor the teeth and urine.

Extreme effects on the skin have prompted the dramatic term "photomutilation". For people with this condition, it is crucial to avoid sunlight and to take a vitamin D supplement. Pending advances in gene therapies, the definitive treatment is allogeneic bone marrow stem cell transplantation.

To conclude our consideration of the porphyrias, I would emphasize that these are extremely important disorders for individuals and their families affected by them. Rare as they are, these conditions deserve greater physician and public awareness. Caregivers, however, would face too high a hurdle if they were expected to have the entire heme synthetic pathway and precursor-induced syndromes memorized: it is sufficient that some combination of neurovisceral symptoms, skin sensitivity to sunlight, discoloration of the urine, and abnormalities in routine blood tests of liver function would trigger consideration of a porphyric disorder. That would then lead to performing appropriate blood and urine tests. Careful interpretation of test results would in turn lead to accuracy in diagnosis (or support the lack of a diagnosis). Patients and family members will always need best advice on supportive care. And the porphyrias are excellent candidates for promising gene therapies.

Further Reading

Karp Leaf R, Dickey AK. How I treat erythropoietic protoporphyria and X-linked protoporphyria. Blood. 2023;141:2921–31. https://doi.org/10.1182/blood.2022018688.

Levy KE. Congenital erythropoietic porphyria. https://www.uptodate.com/contents/congenital-erythropoietic-porphyria.

Mittal S, Anderson KE. Erythropoietic protoporphyria and X-linked protoporphyria. https://www.uptodate.com/contents/erythropoietic-protoporphyria-and-x-linked-protoporphyria.

Mohan G, Madan A. ALA dehydratase deficiency porphyria. https://ncbi.nlm.nih.gov/books/NBK560836/.

Nemeth E, Mathew C. Iron physiology, iron overload, and the porphyrias. In: Altman JK, Mandernach MW, Naik RP, Ulrickson M, editors. ASH-SAP: American Society of Hematology Self-Assessment Program. 8th ed. Washington, DC: American Society of Hematology; 2022. p. 124–35.

Sassa S. The hematologic aspects of porphyria. In: Beutler E, et al., editors. Williams Hematology. 6th ed. New York: McGraw-Hill; 2001. p. 703–20.

Singal AK, Anderson KE. Porphyria cutanea tarda and hepatoerythropoietic porphyria: Pathogenesis, clinical manifestations, and diagnosis. https://www.uptodate.com/contents/porphyria-cutanea-tarda-and-hepatoerythropoietic-porphyria-pathogenesis-clinical-manifestations-and-diagnosis.

Sood GK, Anderson KE. Acute intermittent porphyria: Pathogenesis, clinical features, and diagnosis. https://www.uptodate.com/contents/acute-intermittent-porphyria-pathogenesis-clinical-features-and-diagnosis.

Wiley JS, Moore MR. Heme biosynthesis and its disorders: porphyrias and sideroblastic anemias. In: Hoffman R, et al., editors. Hematology: basic principles and practice. 3rd ed. New York: Churchill Livingstone; 2000. p. 428–45.

Yasuda M, Keel S, Balwani M. RNA interference therapy in acute hepatic porphyrias. Blood. 2023;142:1589–99.

Chapter 11
Hemoglobin Toxins

Monday, September 25, 1944, was a most unusual day at the Beekman-Downtown Hospital in Lower Manhattan, New York City. What happened was chronicled by Berton Roueché, one of the finest medical mystery writers of the twentieth century, in perhaps his most famous story, "The Case of the Eleven Blue Men".[1]

All together, 11 men, starting that morning, became suddenly ill with various degrees of wretching, abdominal pain, altered consciousness, and shock. All of them looked blue. One died. The receiving doctors were nonplussed as to diagnosis. The men who survived were lucky, as they were given no meaningful treatments. The chief epidemiologist for the city was on the case right away. All the men had dined that morning at the nearby Eclipse Cafe, which had been cited for numerous health violations, where the meals were cheap and attracted the down-and-outers. The place was "strictly a horse market, and dirtier than most." The patients were elderly and "delapidated". About 125 customers, including all 11 sick men, had eaten oatmeal that morning into which the cook later admitted having likely but accidentally thrown into the pot a look-alike for table salt, sodium nitrite, used for preserving meats. It happens to be a close taste-alike for table salt (sodium chloride). The amount was probably enough to be toxic, but possibly not enough to do the worst damage. One of 17 salt shakers on the tables also had sodium nitrite, and the investigators surmised that some victims may have further salted their food from that shaker. The blue discoloration of the victims' skin suggested toxic methemoglobinemia, and blood tests confirmed it. Sodium nitrite was ruled the cause. The Eclipse Cafe was immediately closed during the investigation and months later shut down for good.

---

[1] First published in the May 28, 1948, issue of *The New Yorker*.

© The Author(s), under exclusive license to Springer Nature Switzerland AG 2024
M. H. Rosove, *Life's Blood*, https://doi.org/10.1007/978-3-031-61150-6_11

Toxic Methemoglobinemia—*Blue* Skin and *Chocolate* Blood

Methemoglobinemia, you may recall, is a condition in which the iron in heme has been oxidized to the ferric (Fe^{3+}) state, useless for delivering oxygen. The congenital methemoglobinemias described in Chap. 9 usually produce only a cosmetic problem—the telltale bluish cast to the skin when methemoglobin exceeds 1.5 g/dL in the blood, or roughly 10–12% of the total. Toxic methemoglobinemia is another story, because it can reach life-threatening extremes. Not only does heme in the Fe^{3+} state not bind oxygen, but even one oxidized heme in the hemoglobin tetramer increases the oxygen affinity of the remaining hemes, resulting in a left shift and reluctance to release oxygen to the tissues. And while inheriting two abnormal genes for cytochrome b5 reductase (Cyb5R) is necessary to manifest congenital methemoglobinemia, inheriting just one is enough to increase the seriousness of methemoglobinemia from any toxic cause. The same applies to glucose-6-phosphate dehydrogenase (G6PD) deficiency.

Numerous environmental chemicals and commonly used drugs have been implicated in causing toxic methemoglobinemia. Some are well-known, and physicians must be aware of them although they only seldom or rarely cause a problem.[2] The diagnosis is simple and straightforward from readily available blood testing. In the worst cases, the blood appears chocolate brown, practically diagnostic by itself.

Methemoglobinemia is likely to produce symptoms when it is greater than 10–20% of the total hemoglobin, and it is a medical emergency when greater than about 30%. Methylene blue given intravenously is the most rapidly effective treatment. The drug directly reduces Fe^{3+} back to Fe^{2+}, and benefit is evident within minutes to an hour. Emergency room physicians and hematologists have a saying: "Blue for the blue." A danger is that methylene blue paradoxically causes red cell damage if G6PD is deficient, which is not infrequently the case in descendants from populations in the malaria zone—and G6PD-deficient individuals are predisposed to toxic methemoglobinemia in the first place. Therefore, if an individual's geographical or family background point to possible deficiency, methylene blue has to be considered a risk—G6PD test results take time, and there is no time in an emergency. In such cases, physicians may try low-dose methylene blue to test a patient's tolerance. High-dose intravenous vitamin C is a slower-acting and distant-second-place choice. Hyperbaric oxygen, blood transfusion, or whole blood exchange transfusion are important options in extreme cases.

---

[2] The most important include dapsone (used for a number of medical conditions), benzocaine (a topical anesthetic for endoscopic and bronchoscopic procedures, causing methemoglobinemia in about 1 in 1500 to 3000 people), phenazopyridine (an oral urinary tract anesthetic), rasburicase (to stop uric acid production in certain oncologic emergencies, but which generates hydrogen peroxide, an oxidant), inhaled nitric oxide (used therapeutically for severe pulmonary hypertension), recreational amyl nitrite, sodium nitrite, well water (containing sodium nitrate that the bacterial flora of the gastrointestinal tract converts to nitrite), anti-malarial drugs (chloroquine, hydroxychloroquine, primaquine, tafenoquine), and some sulfa drugs.

Sulfhemoglobinemia—*Blue* Skin and *Green* Blood

Another type of toxic damage to hemoglobin is sulfhemoglobinemia. Any drug or industrial compound capable of donating a sulfur atom can be the cause,[3] but even so, sulfhemoglobinemia is exceptionally rare. As little as 0.5 g/dL sulfhemoglobin produces a bluish discoloration of the skin, and the blood has a bizarre bluish-green to greenish-black color. As with methemoglobinemia, there is a similar discordance between arterial partial oxygen pressure and oxygen saturation. Diagnosis is similarly straightforward by the same methods. The unwanted sulfur atom that binds to heme does not oxidize the ferrous (Fe^{2+}) iron, but nevertheless renders that heme incapable of binding oxygen. Curiously, however, it reduces the oxygen affinity of the remaining unsulfurated hemes in the hemoglobin tetramer. The hemoglobin-oxygen dissociation curve is thus shifted markedly rightward, the opposite of methemoglobin. With that, sulhemoglobin readily gives up oxygen to tissues. Thus the condition is not concerning unless the degree of sulfhemoglobinemia is extreme. (Oxygen loading onto hemoglobin in the lungs would be compromised if the problem were to occur at marked altitude or in someone with serious lung disease.) Even though sulfhemoglobin is irreversible and there is no specific treatment, the problem disappears when the offending toxin has been eliminated and the abnormal red cells are recycled at the end of their life spans. Red cell or exchange transfusion are seldom needed.

Hydrogen Cyanide—*Cherry Red* Skin and *Bright Red* Venous Blood

Hydrogen cyanide (chemical formula HCN) and its salts are so renowned for their lethal properties that the term "cyanide" is practically synonymous with "deadly poison". Poisoning can occur via inhalation, swallowing, or through skin contact. Depending on the amount and route of exposure, cyanide can render a person unconsciousness within seconds and dead within a minute or two, or at the other extreme may cause symptoms but with eventual recovery. History is replete with famous examples of deliberate poisoning, including the Zyklon B gas used by the Nazis to exterminate Jews during the Holocaust, suicides (Eva Braun, Heinrich Himmler, Hermann Göring), murders (by Joseph and Magda Goebbels of their six children), mass murder-suicide (Jonestown), and capital punishment (California's historical San Quentin State Prison gas chamber, last employed in 1993, dismantled in 2019).

---

[3] Thus sulfonamides and derivatives are especially suspect causes, most of the same causes of toxic methemoglobinemia, as well as metoclopramide, phenacetin, acetanilid, sulfasalazine, phenazopyridine, sumatriptan, and hydrogen sulfide.

More commonly, people come in contact with cyanide through building fires, which generate it along with toxic oxides of nitrogen, and carbon monoxide. Patients being treated in the hospital for severe hypertensive emergencies with sodium nitroprusside (now seldom used) have to be monitored for cyanide toxicity, as the drug generates not just nitric oxide as intended to dilate blood vessels but also cyanide. Raw or improperly prepared cassava and bamboo shoots are toxic, and the seeds of various fruits such as apricots, greengages, red and black cherries, apples, peaches, and plums contain amygdalin that generates cyanide. Fortunately, the quantity of seeds necessary to cause trouble is quite large—and most people avoid and are not interested in consuming such seeds anyway.

Cyanide inhibits cell metabolism throughout the body by interrupting the operation of a complex protein—mitochondrial cytochrome c oxidase, called COX—that incorporates heme and copper. COX is ubiquitous in nature from primitive bacteria through vertebrates, serving electron transfer as the heme moiety flips between Fe^{3+} and Fe^{2+} states. By means of this electron transfer, ATP is generated for oxygen-dependent cellular respiration and energy production. Cyanide takes advantage of COX by binding to it at the moment its heme is in the ferric (Fe^{3+}) state. That prevents iron from toggling between the two oxidation states. Electron transport is interrupted, cellular respiration stops, and oxygen, although inhaled and transported, is dead-ended before it can be used. Venous blood thus looks as bright red as arterial blood, and the skin may be cherry red. The diagnosis of cyanide poisoning is difficult unless a source is obvious or has been divulged, and even then, therapeutic interventions may be too late; or neurologic damage may be permanent.

An antidote for cyanide poisoning is methemoglobin. Its heme in the Fe^{3+} state can compete for cyanide with the Fe^{3+} in COX. When methemoglobin binds with cyanide, it become innocuous cyanomethemoglobin and is cleared. Thus among the emergency treatments for cyanide poisoning is inhaled amyl nitrite or intravenous sodium nitrite to deliberately produce methemoglobin: in this situation methemoglobin is properly seen as therapeutic, not toxic. It may seem bizarre that one would induce one disorder to treat another, but that is the case here, as the small amount of methemoglobin that normally circulates, about 1% of the total hemoglobin, is quite obviously insufficient to protect against toxic-range cyanide exposure. Hydroxocobalamin and sodium thiosulfate are also part of the treatment of acute cyanide toxicity.

Hydrogen Sulfide

Like cyanide, hydrogen sulfide (chemical formula H_2S) and its salts are toxic, also by inhibiting COX. And as with cyanide, one of the emergency treatments is inhaled amyl nitrite or intravenous sodium nitrite; methemoglobin competes with COX for sulfide, producing in this case sulfhemoglobin that is relatively innocuous. Sulfide poisoning is virtually always accidental.

Sulfur is an essential element in organic chemical processes. The smell of H_2S is familiar, that of rotten eggs, swamps, and sewage. The gas is emitted from landfills, petroleum refineries, paper mills, methane deposits, flatus, and well water—wherever bacteria break down organic material in the relative absence of oxygen. Hydrogen sulfide is also emitted inorganically from volcanic vents and hot springs. Our sense of smell can detect sulfide at even 0.001 parts per million (ppm); likely this sensitivity evolved for its survival value since escaping from the obnoxious odor is a natural reaction. However, at 100 ppm, our olfactory sensors are overwhelmed, and at 1000 ppm H_2S is lethal. Of historical interest is that the chemical formula of hydrogen sulfide was discovered in 1777 by Carl Wilhelm Scheele, the Swedish-German chemist who described oxygen.

Carbon Monoxide—*Bright Red* Venous Blood

Among the hemoglobin toxins, carbon monoxide (chemical formula CO) is in a class by itself as a direct and potent hemoglobin toxin, and it is without question the one that poses the greatest individual and societal concern. In the United States, carbon monoxide toxicity results in up to 50,000 emergency room visits and about 400–500 deaths annually. Most are accidental, some are by suicide, and some rarely by homicide. Carbon monoxide was the poison used by the Nazis from about 1939–1941 in stationary killing centers and sealed "gas vans", to murder disabled citizens they deemed physically and mentally unfit or racially inferior—and Jews only for being Jews.

Tasteless, odorless, non-irritating, and invisible, carbon monoxide also does not produce a sense of shortness of breath, even as one is about to lose consciousness and die, making it an exceptionally stealthy and dangerous environmental hazard. Its deadly effects were undoubtedly known to our ancient human forebears in Africa and the Middle East, who for a million years or more were using fire in hearths, as evidenced by remnant burnt stones and wood fragments. Fire provided light, warmth, a means of cooking food, and protection from predators, and experience would have shown those who built fires inside caves that fumes were toxic. They presumably took precautions, imperfect as they might have been.

It was not until 1772 that Joseph Priestley synthesized CO and Scheele isolated it from charcoal, suggesting that CO was what made its fumes poisonous. CO is the product of incomplete combustion when there is too little oxygen to allow a complete conversion to CO_2. It is a fact of nature. Carbon monoxide is emitted naturally during the photochemical degradation of plant material, during forest fires, and from volcanoes. It may accumulate in confined areas of poor ventilation in coal mines, commonly from internal combustion engines and explosives used therein but also from spontaneous oxidation of coal. Stale air has been called "whitedamp", the silent killer of miners. The expression "canary in the coal mine" hearkens back over a century when air quality in a mine shaft was tested by whether a caged canary survived before humans would venture forward. CO is such a fact of nature that it is one of the commoner diatomic molecules in interstellar space after molecular hydrogen (H_2); it is even a component of Halley's Comet and the atmospheres of Venus and Neptune's moon Triton.

Carbon monoxide may be produced from the burning of gasoline, methane (natural gas), wood, charcoal, and other fossil fuels; and also in home and building fires. Most poisonings occur during the cold winter months when heat is generated inside homes or other enclosed spaces from sources improperly vented, from automobile engines left running in place, and in the interiors of motorized boats and yachts.

A famous case of accidental carbon monoxide poisoning involved Richard E. Byrd, one of the most honored and decorated American naval officers of the past century. During the autumn and winter austral months of his second, large-scale Antarctic expedition in 1934, he manned by himself an underground "Advance Base" in the ice at latitude 80° 08' S for meteorological and auroral research, 215 km south of the main base (Little America II), to become the first human to winter south of 80° South. Cold and darkness were extreme, preventing rescue in case of emergency. Two months into his stay, a generator engine's exhaust vent became obstructed with rime. Byrd was instantly overcome by all the symptoms of severe CO poisoning—impaired consciousness, dizziness, profound headache, nausea, and a hammering heart. He correctly self-diagnosed his condition, and knowing CO rises in air, he kept to the floor, shut off the engine, and slowly recovered from his near fatal experience over many weeks until rescue became possible with the approaching spring season.

All the toxic manifestations of carbon monoxide result from tissue oxygen deprivation. Like oxygen, CO binds reversibly to ferrous (Fe^{2+}) iron in heme, but 210 times more avidly; and it dissociates far more slowly. It causes a markedly adverse left shift in the hemoglobin-oxygen dissociation curve—that unwillingness to release oxygen to the tissues—and can bind to myoglobin and interrupt electron transfer in COX.

Hemoglobin under normal circumstances is under 2% carboxyhemoglobin (COHb) from the natural recycling of heme from senescent red cells, higher if there is a condition that shortens red cell survival and thus increased red cell turnover. These low CO levels are inconsequential. But with chronic CO exposure from moderate to heavy smoking, levels can reach 10%, and red cell 2,3-BPG production is suppressed; the oxygen displacement plus left shift results in increased red cell production that contributes to the tendency toward thrombosis already associated with smoking-related vascular disease. COHb levels above 10% usually indicate poisoning. Symptoms are present at 10–20%, and they become severe and life-threatening at 30–40%. Levels above 50% are usually fatal, and it requires a CO concentration in the air of just 0.1% (1000 ppm) to raise blood COHb to about 50%.

Natural atmospheric concentration of CO is 0.1 ppm. In homes it is commonly 0.5–5 ppm and even up to 15 ppm despite properly regulated gas appliances. Home gas and wood fires with faulty vents or chimneys are very hazardous. The exhaust fumes from automobiles even with catalytic converters contain greater than 1000 ppm CO; without catalytic converters, levels reach 30,000–100,000 ppm. The United States Consumer Product Safety Commission recommends that every home have a properly maintained CO monitor with an audio alarm. Many models are available. They are inexpensive, portable, and can be carried to lodgings away from home. (Do you have one?) Most smoke detectors do not detect carbon monoxide, and vice versa.

A clue to the diagnosis of carbon monoxide poisoning is that venous blood is red like arterial blood because the light absorption spectrum of COHb is close to that of oxygenated hemoglobin. For that reason, routinely used pulse oximeters will not differentiate the two, but blood measurement by the technique of co-oximetry in a clinical laboratory will.

Treatment of toxicity involves first removing the victim from the source of CO, then giving high-flow oxygen (100% normobaric oxygen through a non-rebreathing mask or on a ventilator if the victim is in a coma), and treating other possible accompanying toxicities such as cyanide in dwelling fires. If one is breathing only air, CO takes too long to dissipate with a half-life of 250–320 min, but even with high-flow supplemental oxygen it speeds up only to 75–90 min. Oxygen in a hyperbaric chamber at 2.5–3 atmospheres can shorten the CO half-life further to 30 min, and it dissolves meaningful amounts of oxygen into plasma. Hyperbaric oxygen is valuable as emergency treatment when COHb is 25–40% or greater, when lactic acid buildup means tissues are seriously deprived of oxygen, when there is loss of consciousness or when organs are failing. However, a hyperbaric chamber may not be readily available, and there will be concerns about transfer delays, need for close monitoring inside the chamber, and careful decompression. Recovery times from poisoning are often prolonged, as was the case with Richard Byrd, and neuropsychiatric consequences can be permanent. Carbon monoxide that is bound to fetal hemoglobin has an even longer half-life than usual, and therefore oxygen management of a pregnant woman or newborn is necessarily more aggressive.

To close our discussion, we have seen in this chapter that despite hemoglobin's splendid overall design and function, the molecule is vulnerable to chemical compounds that can disturb it, just as it can be affected by genetic mutations that change globin subunit and heme synthesis. Perhaps in these abnormalities one can appreciate a small silver lining: these anomalous circumstances have taught us much about various aspects of hemoglobin structure and function. Despite ever-evolving understandings of hemoglobin, science has yet been unable to duplicate it. The next, last chapter will address the search for a substitute.

Further Reading

[Anon.] Carbon monoxide. https://en.wikipedia.org/wiki/Carbon_monoxide.

[Anon.] Control of fire by early humans. https://en.wikipedia.org/wiki/Control_of_fire_by_early_humans.

[Anon.] Cyanide in fruit seeds: How dangerous is an apple? https://www.theguardian.com/technology/2015/oct/11/cyanide-in-fruit-seeds-how-dangerous-is-an-apple.

[Anon.] Cyanide poisoning. https://en.wikipedia.org/wiki/Cyanide_poisoning.

[Anon.] Cytochrome c oxidase. https://en.wikipedia.org/wiki/Cytochrome_c_oxidase.

Byrd RE. Alone. New York: G. P. Putnam's Sons; 1938.

[CPSC.] CO Alarms. https://www.cpsc.gov/Safety-Education/Safety-Education-Centers/Carbon-Monoxide-Information-Center/CO-Alarms.

Lavones EJ. Carbon monoxide poisoning. In: Haddad and Winchester's clinical management of poisoning and drug overdose. 4th ed. Philadelphia: Saunders/Elsevier; 2007. p. 1297–307. https://doi.org/10.1016/B978-0-7216-0693-4.50092-x.

Manaker S, Perry H. Carbon monoxide poisoning. https://www.uptodate.com/contents/carbon-monoxide-poisoning [update of January 2023].

Park CM, Nagel RL. Sulfhemoglobinemia. Clinical and molecular aspects. N Engl J Med. 1984;310:1579–84. https://doi.org/10.1056/NEJM198406143102407.

Prchal JT. Methemoglobinemia. https://www.uptodate.com/contents/methemoglobinemia (update of 08 November 2022).

Roueché B. The case of the eleven blue men. The New Yorker, 28 May 1948.

Sawaya A, Menezes RG. Hydrogen sulfide toxicity. StatPearls (update of 12 September 2022). https://www.ncbi.nlm.nih.gov/books/NBK559264/.

Chapter 12
Red Blood Cell Transfusion and the Quest for a Hemoglobin Substitute

The quest for what is commonly called a "blood substitute", but which is more accurately a "hemoglobin substitute" or "substitute oxygen transporter", goes back many decades to when battlefield demands during World War II made obvious the need for a transfusion alternative. Blood products remain among the mainstays of support for surgery and trauma, complicated obstetrical deliveries, organ transplants, complex therapies for malignancies, and treatments for other serious illnesses. And demands surge during natural disasters, epidemics, terror and crisis events. But the practice of blood transfusion is not without its complexities and risks, hence the ongoing search for safe and effective, convenient alternatives.

The effort to develop substitutes has been multinational, involving thousands of people in research and development, and vast sums of capital investment. The commitment speaks to how important the matter is on the one hand, and on the other, how valuable the financial promise for anyone who develops a room-temperature, off-the-shelf, stable product. While hemoglobin substitutes have reached human clinical trials—and even received some short-lived approvals for use in the United States and abroad—a hemoglobin substitute that is effective, safe, and practical has remained elusive. Currently there is no substitute approved by the U.S. Food and Drug Administration. Regardless, the quest continues. Until there is a satisfactory product, hemoglobin replacement will remain dependent solely on red cell transfusion from blood given voluntarily by appropriately screened human donors. A brief survey of blood transfusion should help make clear why researchers continue to pursue a hemoglobin substitute.

Red Blood Cell Transfusion

Transfusion dates back at least to the 1500s when Spanish conquerors arriving in the western coastal regions of South America witnessed Incas practicing human-to-human transfusion. In 1665, a British physician named Richard Lower transfused

M. H. Rosove, *Life's Blood*, https://doi.org/10.1007/978-3-031-61150-6_12

blood from one dog to another, and then in 1667 he tried transfusing small amounts of blood from a sheep to a human. This practice at first seemed promising, but when the French physician Jean-Baptiste Denys in 1667 transfused blood from sheep or calves to humans, serious complications or death resulted. Thus the practice was abandoned by 1670. Then, almost a century and a half later in 1818, a British obstetrician James Blundell transfused human blood to a woman with postpartum bleeding, opening an era of blood transfusion that continues to this day.

At the beginning, life-threatening or fatal reactions from human-to-human transfusion were common. The procedure might go smoothly in one case, but in the next, it might be accompanied by abrupt, massive red cell destruction with shock, organ failure, and death, a reaction that has long been called a "major hemolytic transfusion reaction". In 1901, the Austrian-American immunologist Karl Landsteiner discovered the cause: one person's serum (the portion of clotted blood that is not blood cells) might or might not clump another's red cells. Adverse reactions, he found, resulted when the recipient treated the transfused blood as foreign and activated a rejection of it. Red cells, he discovered, fell into four groups: those possessing a so-called A antigen, B antigen, both, or neither. These are blood types A, B, AB, and O (the last named from the German *ohne*, without). An individual with type A would have antibodies to B, and vice versa; type O would make antibodies to both, and type AB to neither. This discovery permitted identifying compatibility between donor and recipient, and it revolutionized transfusion medicine. Landsteiner is often called "the father of transfusion medicine".[1]

The antigens are glycoproteins, and inheritance is Mendelian. Each parent provides a gene for A, B, or neither (O). One A and/or B determines the A, B and AB types. Type O is recessive, meaning neither A nor B was inherited from either parent. The A and B glycoprotein antigens are similar to those found widely in nature including in foods and from micro-organisms. Thus we are exposed early in life, and that is when antibody formation begins.

Landsteiner continued his work on blood group antigens and was co-discoverer of the Rh group in 1937, the most important of which is Rh factor D, abbreviated Rh(D). It is so important that it is part of the nomenclature of one's blood type. If you are "O negative", that means type O and Rh(D) negative. Rh(D) negative blood is especially valuable to blood banks because it is the only type that Rh(D) negative patients can safely receive. About 15% of Caucasians are Rh(D) negative; percentages are lower among Blacks and Asians. Beyond Rh(D), a number of red cell antigen systems can incite antibodies and transfusion reactions, though their effects are less predictable. These include other members of the Rh group and the Kell, Duffy, Kidd, Diego, and MNS groups.

Today, major hemolytic transfusion reactions have all but been eliminated by strict policies starting with blood drawing, blood collection tube labeling, procedures at the blood banks, and finally matching blood and patient at the bedside, all

---

[1] Landsteiner was the Nobel laureate in Physiology or Medicine in 1930.

to be certain of no ABO mismatch. One must not give type A blood to someone with type O or B blood; or type B to someone with O or A; or AB to anyone except AB. The result of a mismatch is dire—abrupt destruction of the transfused red cells, a state of shock, and possible fatal outcome.[2]

Even if the recipient of a transfusion is perfectly ABO and Rh(D) matched to the donor, however, the recipient may face complications on account of lacking an antigen the donor red cells have. If the recipient's immune system "sees" the foreign antigen, several days after transfusion the recipient may generate antibodies against it and destroy the transfused cells. This is a "delayed hemolytic transfusion reaction". Reactions vary in seriousness depending on the antigen in question and other factors.

Individuals who have been previously transfused or have been pregnant may carry temporary or longstanding antibodies. (Antibodies against red cell antigens may rarely result from an unrelated medical illness as well. And they can be passively and unintentionally transferred when an individual is given blood products or gammaglobulin.) Therefore, a recipient's serum is always tested ("crossmatched") with the donor unit to assure safety. However, an antibody may disappear over time. Then, unwitting exposure later to red cells with the antigen in question is likely incite a reaction a few days afterwards, if it did before. Knowledge of that antibody beforehand will help avoid problems. Occasional recipients of many transfusions develop antibodies against multiple red cell antigens. Finding compatible red cell units for them can be very difficult, even impossible, with life-threatening implications.

Transfusion isn't the only circumstance that can bring an individual's blood in contact with antigens that don't match their own. During pregnancy, some fetal blood enters the maternal circulation. The fetus will almost certainly inherit one or more antigens from its father that its mother lacks, and she may develop antibodies that can cross the placenta and cause red cell destruction in the fetus, often discovered only at birth, referred to as "hemolytic disease of the fetus and newborn". This is thankfully uncommon. The mother herself is unaffected, since she lacks the antigen on her own red cells. Diagnosis is not usually difficult, and the condition can usually be managed satisfactorily. Rh(D) is the principal antigen of concern, prospective parents are always tested, and the matter will come up in maternal-fetal management when the mother is Rh(D) negative and the father Rh(D) positive.

---

[2] A very rare occurrence is homozygous absence of the red cell H antigen, the precursor for both A and B antigens. Irrespective of inheritance for A or B, they are not expressed. Such individuals therefore make both anti-A and anti-B, and also anti-H. Even a type O transfusion in this instance is unacceptable as virtually all blood has the H antigen: the result would be a major reaction unless the red cells came from another rare donor lacking the H antigen.

Problems Attendant to Red Cell Transfusion

All of these aforementioned issues keep the search for a hemoglobin substitute alive. And yet there are even more concerns from red cell transfusion. Common ones are allergic or febrile reactions, usually not serious. Others, however, all uncommon to rare, can be life-threatening. Transfusions may overload the circulation, causing "transfusion-associated circulatory overload" (TACO). Transfusion may occasionally be followed within hours by lung injury when activated white cells and immunoglobulins are deposited in the pulmonary circulation, a condition called "transfusion-related acute lung injury" (TRALI). Red cell transfusion contains scant platelets, and on rare occasion these may incite an immunologic reaction against the recipient's own platelets causing a bleeding disorder, "post-transfusion purpura" (PTP).

It is also possible for a unit of blood to become infected during processing or to carry an infectious micro-organism missed in screening; this happens very rarely, but it is always serious. Hepatitis B and hepatitis C have been virtually eliminated as transfusion risks with accurate donor blood testing beginning in the early 1970s for hepatitis B and 1990 for hepatitis C, but until they were, transfusion-related hepatitis was common.

Of interest is that the virus of SARS-CoV-2 (COVID-19) has not been shown to be transmissible through blood transfusion. But worrisome is that donated blood may spread a heretofore unidentified micro-organism that *can* be spread quickly in the human population. That is what happened in 1979 when the first cases of the acquired immunodeficiency syndrome (AIDS) were identified. AIDS was often fatal then, its cause was yet unknown, but within the year blood transfusion was determined to be one of a few vehicles of transmitting the disease, whatever it was. Despite advisories that high-risk individuals not donate blood, no one could guarantee that people would adhere to that guidance, and several years elapsed before the human immunodeficiency virus (HIV) was identified and blood product screening could definitively eliminate the risk. Until 1985, patients dreaded transfusions, and many not irrationally refused them, even when their anemia was extreme. Could this happen again someday? We have to assume it could. Or will.

One final complication of note is "transfusion-associated graft-versus-host disease" (TA-GVHD). It is rare but extremely serious and often fatal. T-lymphocytes in the donor blood attack the recipient's bone marrow causing complete failure of blood cell production, along with attacks on other organs. The Association for Advancement of Blood & Biotherapies advises that blood products be irradiated to destroy the T-lymphocytes if the recipient fits a high-risk category: patients immunocompromised by background condition, medications, or organ transplant status; or when the donor is related to the recipient (including if a member of a highly inbred population).

Given the foregoing, one might wonder why a patient would ever agree to accept transfusion under any circumstances. But it is important to realize that overall, adverse reactions are uncommon, and the most serious ones are rare. Transfusion remains a potentially life-saving intervention. The good news is that clinical studies in the last two to three decades have persuaded health care providers that hospitalized patients can tolerate greater degrees of anemia than previously thought. That has both eased the burden on the blood supply and reduced the risk of complications simply by reducing the number of transfusions given.

Yet the critical need remains, along with the demanding process of safely procuring, processing, storing, and testing that insurance systems only marginally compensate. In the U.S., the blood supply is always tenuous, though it is usually sufficient to meet needs. Most volunteers give a "unit" of blood, that is, about 450–500 mL, representing roughly 8–12% of their blood volume. The process is nearly always uneventful with routine and proper prescreening. A small amount of the blood is tested to be as sure as possible there is no transmissible infective agent. An approved unit is separated into its principal components—red cells, plasma, and platelets, and possibly albumin, clotting proteins, and immunoglobulins. Some specialized donors provide only plasma or platelets.

Temporary shortages are common: donations often slow down as the December holiday season approaches, but demand does not; and blood donation centers often plead for volunteers in December and January. Those who donate blood then, or at any time, are modern-day good Samaritans. Demand is usually very high for type O negative blood, the "universal donor", which is always in short supply as only 5% of random donors are that type. O negative blood can be used in emergencies even before the recipient can be confidently blood typed. Those who know they are O negative and donate often, especially in times of societal duress, deserve all our gratitude.

The issues of supply, as well as the problems that still come with red cell transfusion, are the chief rationale for pursuing hemoglobin substitutes.

The Quest for a Hemoglobin Substitute

Two kinds of hemoglobin substitutes have come under investigation over the last half century or so: modified hemoglobins and perfluorocarbons (PFCs).

Modified hemoglobins. Since native hemoglobin cannot circulate freely on its own outside the red cell without causing trouble, researchers since the late 1970s put an enormous effort into modifying hemoglobin so that perhaps it might be tolerated. They first separated hemoglobin from other red cell elements, then chemically transformed it. They investigated a number of variations, but only one, OxyGlobin,[3] is still used, and only in dogs. It is a bovine-derived hemoglobin approved in the United States and Europe but currently experiencing a production lapse. Other

---

[3] OxyGlobin (originally OPK Biotech, Cambridge, MA; now HbO2 Therapeutics, Souderton, PA, and Groningen, Netherlands.).

engineered hemoglobins underwent extensive testing.[4] They all fell short of their promise—variously because they caused vasoconstriction, high blood pressure and/or compromised cardiac output, possibly by interfering with the nitric oxide cycle; or caused an unacceptable left shift in the hemoglobin-oxygen dissociation curve related to reduced 2,3-BPG effect; or the modified hemoglobin oxidized to methemoglobin. In some cases research funds dried up, or the developing company declared bankruptcy. Interest remains, however, in a modified bovine hemoglobin (Hemopure, HBOC-201, HbO2 Therapeutics) that was introduced into clinical trials over 20 years ago. It is not FDA-approved, but it is currently available through the manufacturer's expanded access program for life-threatening anemia when there is no transfusion option.[5]

In a meta-analysis of modified hemoglobins, Natanson and colleagues in 2008 examined 16 clinical trials they considered eligible for scrutiny because they were prospective and because patients were randomly preselected (like drawing straws) to treatment or not. The studies involved five different hemoglobin substitutes and 3711 patients. There were more deaths in the treated patients (164) than in the untreated ones (123), and there were more myocardial infarctions in the treated patients (59) than in the untreated ones (16). There were no clear associations with the product used or reason for treatment. Given the serious side effects of treatment and questionable benefits, plus concerns over the design of a number of the trials, Fergusson and McIntyre in an editorial accompanying the Natanson paper called for halting use of all the modified hemoglobins that existed at that time.

Perfluorocarbons. The perfluorocarbons (PFCs) are completely different. They are synthetic hydrocarbon chains in which fluorine nearly or completely substitutes for hydrogen. The fluorine-carbon bonds are very strong and stable, rendering PFCs chemically inert. They are not soluble in water and are emulsified for infusion. Half-lives in the circulation vary from hours to days, and they leave by exhalation. Suspended PFC particles range in size from 0.1 to 0.3 μm, far smaller than the average red cell diameter of 7 μm and capillaries whose cross sections are about 5–10 μm: thus PFCs traverse the capillary circulation easily.

They dissolve and release gases readily due to their weak intermolecular interactions; however, they do not actually bind oxygen. Rather, they carry and release it passively in direct proportion to the partial pressure of oxygen: they are thus non-cooperative. A further disadvantage is that they carry oxygen optimally only at arterial partial pressures of oxygen well above what can be achieved from ambient air. Thus recipients had to be breathing a high concentration of oxygen. It follows that PFCs would not perform in serious lung disease when even high-flow oxygen might not provide high oxygen tension reaching the arteries.

---

[4] HemAssist (Baxter), Hemolink (Hemosol, Inc.), Hemopure (Biopure Corp.), PolyHeme (Northfield Laboratories), Dextran-Hemoglobin (Dextro-Sang Corp.), Hemotech (HemoBiotech), Optro (Somatogen), PHP (Somatogen), PHP (Apex Biosciences), and Hemospan (Sangart).

[5] The side effect profile is similar to other modified hemoglobins. The P50 is significantly right shifted to 40 mm Hg. Half-life is 19 h. Repeated transfusions lead to iron overload.

Research on PFCs as a hemoglobin substitute began in 1966 when Clark and Gollan showed that mice and cats could survive breathing PFCs, but all the animals experienced lung toxicity. A number of engineered PFCs were developed for intravenous use instead and underwent clinical trials, but none lasted.[6] The only one that ever achieved FDA approval in the U.S. (as well as approval in a few other countries) was Fluosol-DA. It was introduced into clinical trials in the U.S. in 1979 and was welcomed enthusiastically during the handful of years in the era when AIDS could be transmitted by transfusion. But Fluosol-DA could cause asthmatic reactions and pulmonary complications from overloading the circulation. As patients had to be given very high flow oxygen continuously, they were at risk of lung toxicity from prolonged high-flow oxygen use itself. Fluosol-DA adversely activated the immune system, and some of it was retained in the liver and spleen raising concern whether that might have adverse consequences. It was also inconvenient: it had to be stored frozen and thawed before use.

Gould and colleagues' report of Fluosol-DA in 1986 in *The New England Journal of Medicine* suggested that it had limited if any value. They studied 23 patients undergoing surgery who had indicated beforehand they would refuse blood transfusion on religious grounds. (Concern for such patients, including Jehovah's Witnesses who are forbidden to accept red cell transfusion, and members of other religious groups and individuals who similarly reject it, has been a driver of research.) To be eligible in the study to receive Fluosol-DA, the post-operative hemoglobin level had to be 3.5 g/dL or less, or the oxygen extraction percentage from hemoglobin between arterial and mixed venous blood had to be over 50% (normal at rest being about 28%). Either of these threshold values would be extremely compelling for lifesaving transfusion because they meant oxygen transport and delivery were critically inadequate. Among the 23 patients considered for Fluosol-DA, 15 did not qualify, their lowest mean hemoglobin value was 7.2 g/dL, and 14 of them survived. Eight did qualify, all were given Fluosol-DA in optimal dosage including inhaled oxygen at high partial pressure, but the increase in oxygen carrying capacity of blood proved to be disappointing, though no treated patient had side effects. Their mean lowest hemoglobin level was 1.8 g/dL. Six of the eight died; one survived after a court order forced him to be transfused against his wishes. Gould and co-authors concluded, "Fluosol-DA is unnecessary when anemia is moderate and ineffective when it is severe." The FDA withdrew the approval in 1994 for various reasons including side effects and difficulties around handling the frozen material.

Current and future directions. The quest for blood substitutes continues. Under the U.S. Department of Defense, the Defense Advanced Research Project Agency (DARPA) early in 2023 funded $46.4 million for development of synthetic blood. Twelve universities and laboratories, led by the University of Maryland,

---

[6] These included Oxyfluor (HemaGen, Sunset Hills, MO), Oxygent (Alliance Pharmaceuticals, Chippenham, Wiltshire, UK), OFNE (Integra Life Sciences, Princeton, NJ), Oxycyte (Aurum Biosciences, Glasgow, Scotland, UK), Perftorin (Vidaphor) (Russia), and Fluosol-DA (Green Cross, Osaka, Japan). Perftorin was approved in Russia in 1994 and in Mexico from 2005 to 2010, but experienced "manufacturing concerns and economic issues".

Baltimore, are working to create the individual components of blood—red cells (with their hemoglobin), plasma, and platelets. The products would theoretically be recombined in specific proportions depending upon need, and removal of antigens would render the components immunologically silent—anyone could receive them. Human trials are anticipated for 2028–2030. While the prospect of realizing the project's goals is daunting, optimism prevails. Other current research programs are examining protein-based non-hemoglobin oxygen carriers, nanoparticle carriers, hemerythrin as a surrogate transporter, and hemoglobin shielded within phospholipid vesicles.

The ongoing pursuit of a hemoglobin substitute prompts two thoughts. First, the quest for a worthy substitute for life's blood is as strong as it was over half a century ago, and for good reasons. May the effort continue. Success would be an epic achievement. And second, our failure to date to produce a substitute, despite our best minds and technologies, must cause us to pause and pay homage to the half billion years of evolution that created red blood cells and the oxygen-carrying specialist we already have: the complex, miraculous molecule we call hemoglobin.

Further Reading

Association for Advancement of Blood & Biotherapies. Irradiation of blood products. https://www.aabb.org/regulatory-and-advocacy/regulatory-affairs/regulatory-for-blood/irradiation.

[Anon.] Blood substitute. https://en.wikipedia.org/wiki/Blood_substitute.

[Anon.] Fluosol. https://www.sciencedirect.com/topics/medicine-and-dentistry/fluosol.

[Anon.] Jehovah's Witnesses and blood transfusion. https://en.wikipedia.org/wiki/Jehovah's_Witnesses_and_blood_transfusions.

[Anon.] OxyGlobin. https://www.drugs.com/vet/oxyglobin-solution.html.

[Anon.] Perfluorocarbon. https://en.wikipedia.org/wiki/Perfluorocarbon.

[Anon.] Perfluorocarbon emulsions. Oxygen therapeutics. https://en.wikipedia.org/wiki/Perfluorocarbon_emulsions#Oxygen_Therapeutics.

Asuma H, Amano T, Kamiyama N, et al. First-in-human phase 1 trial of hemoglobin vesicles as artificial red blood cells developed for use as a transfusion alternative. Blood Adv. 2022;6:5711–5. https://doi.org/10.1182/bloodadvances.2022007977.

Chen J-Y, Scerbo M, Kramer G. A review of blood substitutes: examining the history, clinical trial results, and ethics of hemoglobin-based oxygen carriers. Clinics (Sao Paulo). 2009;64:803–13. https://doi.org/10.1590/S1807-59322009000800016.

Clark LC, Gollan F. Survival of mammals breathing organic liquids equilibrated with oxygen at atmospheric pressure. Science. 1966;152:1755–6.

Dean L. Blood groups and red cell antigens. Bethesda: National Center for Biotechnology Information (US); 2005. Chapter 4, Hemolytic disease of the newborn. https://ncbi.nlm.nih.gov/books/NBK2266

Edwards L. Artificial blood developed for the battlefield. https://medicalxpress.com/news/2010-07-artificial-blood-battlefield.html.

Fergusson DA, McIntyre L. The future of clinical trials evaluating blood substitutes. Am Med Assoc J. 2008;299:2324–6. https://doi.org/10.1001/jama.299.19.jrv80027.

Gould SA, Rosen AL, Sehgal LR, et al. Fluosol-DA as a red-cell substitute in acute anemia. N Engl J Med. 1986;314:1653–6. https://doi.org/10.1056/NEJM198606263142601.

Jahr JS, Guinn NR, Lowery DR, Shore-Lesserson L, Shander A. Blood substitutes and oxygen therapeutics: a review. Anesth Analg. 2021;132:119–29.

Khan F, Singh K, Friedman MT. Artificial blood: the history and current perspectives of blood substitutes. Discover. 2020;8(1):e104, 1–15. https://doi.org/10.15190/d.2020.1.

Kopolovic I, Ostro J, Tsubota H, et al. A systematic review of transfusion-associated graft-versus-host disease. Blood. 2015;126:406–14. https://doi.org/10.1182/blood-2015-01-620872.

Latson GW. Perftoran: history, clinical trials, and pathway forward. In: Blood substitutes and pathway forward. Cham: Springer; 2022. p. 361–7.

Li C, Georgakopoulou A, Newby GA, et al. In vivo HSC prime editing rescues sickle cell disease in a mouse model. Blood. 2023;141:2085–99.

Mer M, Hodgson E, Wallis L, Jacobson B, Levien L, Snyman J, et al. Hemoglobin glutamer-250 (bovine) in South Africa: consensus usage guidelines from clinician experts who have treated patients. Transfusion. 2016;56:2631–6. https://doi.org/10.1111/trf.13726.

Natanson C, Kern SJ, Luri P, Banks SM, Wolfe SM. Cell-free hemoglobin-based blood substitutes and risk of myocardial infarction and death: a meta-analysis. Am Med Assoc J. 2008;299:2304–12. https://doi.org/10.1001/jama.299.19.jrv80007.

Shi PA, Luchsinger LL, Greally JM, Delaney CS. Umbilical cord blood: an undervalued and underutilized resource in allogeneic hematopoietic stem cell transplant and novel cell therapy applications. Curr Opin Hematol. 2022;29:317–26.

Webster H. DARPA puts $46.4 million toward synthetic blood development. https://www.stripes.com/theaters/us/2023-02-02/darpa-46-million-synthetic-blood-development-9018199.html.

Zumberg M, Gorlin J, Griffiths EA, Schwartz G, Fletcher BS, Walsh K, et al. A case study of 10 patients administered HBOC-201 in high doses over a prolonged period: outcomes during severe anemia when transfusion is not an option. Transfusion. 2020;60:932–9. https://doi.org/10.1111/trf.15778.

Index

A
adenosine triphosphate (ATP), 34, 44
adult hemoglobin, *see* Hemoglobin A
Al-Nafis, Ibn, 8–10
altitude, high
adaptation, 81–92
Andes, Tibetan plateau, Mt.
Everest, 86, 88
in humans, birds, mammals, 89
environmental challenges, 82, 84,
86, 88, 89
hemoglobin-oxygen dissociation curve,
84, 87–91
physiological challenges, 81, 83
illnesses, serious, 85
chronic mountain sickness (CMS),
85, 87, 89
high-altitude pulmonary edema
(HAPE), 84
high-altitude cerebral edema
(HACE), 84
Aristotle, 7, 10
Ascaris lumbricoides, 36
Ashby, Winifred, 45
atmosphere, origin and composition, 2, 33, 34
Avery, Oswald, 58

B
Bart's hydrops fetalis syndrome, 63
Bernal, John Desmond, 14
Bert, Paul, 82
Boyle, Robert, 2
blood circulation, 8
increased as compensation, 29

bilirubin, 45
biliverdin, 45
blood doping, 92
blutfarbstof, 13
Braunitzer, Gerhard, 15

C
Caventou, Joseph Bienaimé, 53
carbon monoxide, 45, 107–109
carboxyhemoglobin (COHb), 45
from hemoglobin degradation, 45
poisoning, prevention and
treatment, 105–109
chlorocruorin, 37
combustion and respiration, similarity of, 1, 5
Cooley, Thomas B., 64
Cooley's anemia, 64
copper, 33, 34, 38, 106
in hemocyanin, 37–38
in cytochrome c oxidase (COX), 106–108
oxidation states, 34
Countess of Cinchón, 53
Crick, Francis, 58
cyanide toxicity, causes and treatment, 106
cyanobacteria, 33
cyanotic congenital heart disease, 87
cytochrome c oxidase (COX), 106–108
cytoglobin evolution, in jawless fishes, 35

D
"decompensated erythrocytosis", 87
Denisovans, 86
Diggs, Lemuel W., 74